WHAT ARE THE CHANCES?

What Are the Chances?

Voodoo Deaths, Office Gossip, and Other Adventures in Probability

Bart K. Holland

Printed in the United States of America on acid-free paper
9 8 7 6 5 4 3 2 1

The Johns Hopkins University Press
2715 North Charles Street
Baltimore, Maryland 21218-4363
www.press.jhu.edu

Library of Congress Cataloging-in-Publication Data

Holland, Bart K.
What are the chances? : voodoo deaths, office gossip, and other adventures in probability / Bart K. Holland.
p. cm.
Includes index.
ISBN 0-8018-6941-2 (hardcover : alk. paper)
1. Probabilities—Popular works. 2. Chance—Popular works. I. Title.
QA273.15.H65 2002
519.2–dc21

2001005687

A catalog record for this book is available from the British Library.

For Alicia and Charlie

Contents

Preface

I have never been able to figure out what subject I would most like to study, and I still can't decide. An odd confession perhaps, coming from one with a tenured professorship in a specialized area (and in his fifth decade of life, to boot). Yet, even within the confines of academia, there is such a wide variety of potentially useful and interesting work to choose from that it has always seemed rather arbitrary and limiting to focus on a single subject to the exclusion of others. On the other hand, specialization is required by the job market. No one moves up who remains a "jack of all trades and master of none." And specialization is required in order to be productive in research.

For the scientifically inclined, one solution to this dilemma is to be well trained in the methods of probability and statistics, which can then be applied to a broad range of subjects. I came to this solution neither deliberately nor immediately, but gradually and by accident. In college I took many enlightening courses in anthropology, followed by demography. Taking them in this order, I saw with fascination how regularities emerged from the most individualistic behaviors when societies were studied en masse. Then I saw that the same was true in biology: the laws of probability could summarize and predict phenomena as diverse as population genetics and reactions to medicines. Economics, physics, geology, psychology—everywhere you looked probability mathematics was a useful tool and contributed something important to human understanding.

These days, I use probability and statistics primarily in medical research, to provide physicians with useful information from experiments that would otherwise seem difficult to interpret, if not chaotic. However, I never lost my interest in other areas and have always made a note of interesting (and preferably odd) examples showing probability's applications more generally. Thus this book: I've finally had the opportunity to put my favorite examples on paper, thanks to Trevor Lipscombe and the Johns Hopkins University Press. I hope that you find the examples intriguing, and an effective way to learn some basic concepts of probability and statistical methods. And perhaps you will even share my amazement when you see the extent to which probability influences so many facets of human life.

Acknowledgments

My editor, Trevor C. Lipscombe, D.Phil., was the first to suggest that I write a book such as this, using diverse historical, medical, and everyday examples to illustrate the principles of probability and statistics. (Thus, if you do not like this book you should blame him, not me.) As luck would have it, we share many interests, so we had ample opportunity to trade anecdotes from our collections and to discuss how they could fit in here. We also have similar ideas about the use of language, a similarity that made it easy for me to accept his deft editorial suggestions.

My most loyal reader is Jean M. Donahue, Ph.D., and I am grateful for her insightful review of these pages. Scientist, teacher, muse, and wife, she also shares my work and my love of books; in this I am indeed fortunate. Her parents, Nora and Charles L. Donahue, indulge my writing habit by taking care of our children when I feel the urge to take up the pen, so they certainly contributed directly to making this book possible.

I should also acknowledge John Bogden, Ph.D., and William Halperin, M.D., Dr.P.H., successive chairmen of my department at New Jersey Medical School, for their encouragement. In addition, my colleague Joan Skurnick, Ph.D., provided a number of useful suggestions and the interesting story of her being caught in a whiteout, which inspired an example used in the chapter on random walks.

Finally, I thank NJ Transit, since being delayed on their trains allowed me time to write; family friend Elizabeth R. Nesbitt for naming Esmeralda; and the staff of the George F. Smith Library for obtaining many obscure references.

WHAT ARE THE CHANCES?

1

Roulette Wheels and the Plague

The Role of Probabilities in Prediction

It was gruesome. Corpses—swollen, blue-black, and stinking of decay—were hurled through the air, propelled by a catapult on a trajectory that landed them inside the walls of the besieged city. The city was Caffa, in the Crimea, called Feodosiya in present-day Ukraine. In the fourteenth century, the city was a stronghold of Genoese merchants. It was under attack by Mongol armies, as it had been several times before. An attack in 1344 had shown the city to be nearly impregnable, but it was now two years later and something was different: this time the bubonic plague accompanied the armies from Central Asia. The invading Tartar soldiers were being decimated by the disease, and they also faced an acute sanitation problem caused by the accumulation of dead bodies. Military genius came to their aid. The Mongols had brought along a kind of strong catapult called the *trebuchet*; it was ordinarily used to hurl heavy loads of stone to destroy defenses such as masonry walls and towers. Now, not stones but human missiles rained down upon those behind the walls. An eyewitness by the name of Gabriel de Mussis wrote in a Latin manuscript (still legible today) that the mountains of dead were soon joined by many of the Christian defenders, while those who were able to escape fled the stench and the disease.

The story of Caffa is not just an early case of germ warfare. Some historians and epidemiologists believe that this particular battle marks the starting point of the plague's invasion from Central Asia into Europe. The Genoese who fled to Europe may have brought the bacteria back in rats on their ships, and in the rats' fleas (which hop off and bite people, thus transmitting the *Yersinia pestis* bacteria to the human bloodstream). Whatever the original source, the great European plague of 1348 certainly emanated from Mediterranean port cities. From accounts written by monks and from parish death records, we know it went on to kill somewhere between 25 and 50% of the European population. However, the exact route or routes by which the Black Death came to Europe will never be known for certain.

The bubonic plague happened centuries ago, but the questions it posed then are still with us today. Why do epidemics "break out"? Why can't scientists predict the size, location, and timing of the next outbreak of an "old" disease, such as influenza or measles, much less the coming of "new" diseases, such as AIDS? The difficulty lies in creating accurate scientific models of infection, rather like predicting the weather relies on decent models of the atmospheres and oceans. Epidemics occur when certain chains of events occur; each event has a certain probability of occurring and, as a consequence, an average or expected frequency of occurrence. To predict epidemics, we need to have accurate mathematical models of the process. Modeling involves knowing the steps in the chain and the probability of each one. Then the expected outcome of the whole series of steps can be estimated, in essence by multiplying together all the probabilities and numbers of people involved in a particular scenario. To take a simple example, consider the case in which a disease is spread by person-to-person transmission, let's say by sneezing, as with influenza. An epidemic can continue when each infected person in the population, on average, meets and infects a healthy person, known to epidemiologists as a *susceptible.* There is a certain probability of this happening, as there is for the infection of any given number of new susceptibles. An epidemic can grow when each person meets and infects on average more than one susceptible, and then each of the newly infected persons meets and infects several more in turn, and so on. An epidemic will die out if infected people average less than one "successful" new infection apiece, and a simple chain model of probabilities will reflect this.

A closely related example comes not from the world of medicine, but of comedy. Suppose you make up a joke and tell it to a couple of friends. If it bombs, the joke won't spread. But if your friends laugh, and each of them tells 2 people within 24 hours of hearing it, then the number of people who will have heard it after 24 hours is 2; at the same rate, after 48 hours it will be up to 4, after 3 days it will be 8, and by the end of the week 128 people will have heard your joke. That may sound impressive, but just wait—by the end of the second week, more than 16,000 people will have heard it, and by the end of the month, that number will have reached some 250 million (roughly the population of the United States). Or will they have heard it? How many times have you started to tell a joke only to meet with an "I've heard it" or "That's not funny"? The fact that a population is not infinite, and that some people are immune to a disease or a joke, severely restricts the outcomes of the mathematical models we construct to explain how a joke, or a disease, spreads. If this weren't the case, then we'd have good news and bad news: the good news would be that we could all make our fortunes from chain letters; the bad news would be that the entire population of Europe would have been wiped out by the bubonic plague.

The way that gossip spreads throughout the workplace is another interesting example of this chain reaction mechanism. Gossip, though, has something more profoundly in common with the spread of disease. You hear a juicy piece of information and pass it on to a couple of trusted confidants, they may repeat it as well. After a few person-to-person transmissions, the message is rarely the same as at the beginning. You hear something interesting about Craig and Maureen and tell someone else. When you hear a spicy story a month later about Greg and Noreen, will you recognize it as a garbled version of the original, or as a hot news item to be e-mailed to your friends right away? In the language of genetics, the gossip has mutated and, like a disease that mutates, it can then reinfect someone who had caught the original disease. This happens with influenza, for which the vaccine has to be reformulated each year in order to be effective against newly appearing strains of the disease.

Chain Reactions in Atoms and People

Chains of events are used to understand many important processes in the sciences. The same model—consisting of successive probabilities and producing estimates of overall outcomes—governs the chain reaction in nuclear physics. A chain reaction can be sustained when the atoms in a radioactive substance emit particles, and each atom's particles split (on average) one atom in turn, causing further particle emission, and so on. If more than one atom gets split by particles from the previous atom, then the chain reaction takes off. In nature, radioactive decay occurs and particles are emitted, but chain reactions do not occur: they die out, because the radioactive forms (radioisotopes) of elements are not present in a great concentration (and the nonradioactive isotopes do not "emit and split" the way the radioactive ones do). Thus, the particles emitted, on average, travel harmlessly through the substance without splitting another atom, without reproducing or enhancing their emission. These emitted particles can be very important even when they are not being used to provoke a chain reaction. Radioisotopes have been put to great use in medicine, because certain radioactive materials are attracted to particular bones, organs, or tissues when injected or ingested. Films or radiation-sensitive devices can then be placed next to the body, and the emitted particles produce images that allow diagnosis of cancers and other diseases. In addition, some cancers are treated by radioisotopes because the radiation kills the cancer cells, and the differential absorption by the targeted tissue is a desirable property.

One of the key tasks of the Manhattan Project of the U.S. military in World War II, under the leadership of General Leslie Groves, was to determine how to produce growing chain reactions, in order to make the atomic bomb. The scientists working on the project achieved this by isolating and concentrating radioactive isotopes, which increased the probabilities of particle emission and subsequent nuclear fission throughout the chain. It was also necessary to know, for example, the diameters of atomic nuclei and how to compress atoms closer together, to estimate and improve the chances of every emitted particle splitting a subsequent nucleus, causing further emission and enhanced release of energy. It took some of the finest physicists in the world, including J. Robert Oppenheimer, to accomplish this. On July 16, 1945, a little before 5:30 A.M., the first nuclear weapon was exploded at Los Alamos, New Mexico. Within a month, "Fat Man" and "Little Boy" would be dropped on Japan.

There's a strong analogy here: the mathematics that governs chain reactions is the same as that governing the course of epidemics, because each person or atom must affect the next one in order for sustained transmission to occur. Early on in human history, large epidemics rarely, if ever, happened. Human populations were sparse in prehistoric times when people were all hunters and gatherers. A small band consisting of a few families might all be stricken if one member encountered the measles virus, but odds were low of meeting and infecting other bands during the illness. People infected with communicable diseases could not, on average, meet and infect one susceptible. Epidemics therefore tended to die out quickly, and there is evidence that such diseases were quite rare at first but became more common as the possibilities of transmission rose along with expanding human populations. Paleopathologists see evidence of measles in human remains dating from about 4000 B.C.E. in the area of the Tigris and Euphrates valleys, because the area was dotted then with small cities made possible by the dawn of agriculture. From pathologists' examination of mummies, we know that tuberculosis was known in ancient Egypt from the time when cattle were domesticated and herded in proximity to people, around 1000 B.C.E., and that the disease probably entered human society from the bovine reservoir in which it was prevalent. It takes a large settled society to ensure optimal conditions for person-to-person spread of epidemics: in larger groups there is a greater chance that *someone* within the population is available as a source of infection, and chances are higher that he or she will meet some member of a constantly renewed source of susceptibles (provided by births or immigration). It takes a village to make an epidemic.

For any given population size, the fraction of people who are susceptible to a disease is a key influence on the ultimate size of an epidemic. Smallpox was not an indigenous disease among the tribes native to North America, so none had the immunity conferred by the experience of even the mildest smallpox infection, and almost all were susceptible when colonists arrived from Europe. Many Europeans had immunity as a result of exposure, whereas others among them had active cases of the disease. In 1763, a series of letters between Sir Jeffrey Amherst, British commander in chief for North America, and Colonel Henry Bouquet, in charge of military operations at the Pennsylvania frontier, outlined a plan to provide materials infected with smallpox to the native "Indian" tribes. Once again, it was a plan emanating from an army facing both an enemy and an outbreak. Bouquet reported that the natives were "laying waste to the settlements, destroying the harvest, and butchering men, women, and children." Also, his troops were suffering from an outbreak of smallpox. Bouquet and Amherst agreed that the solution was to attempt to appease (and kill) the enemy with "peace offerings," consisting of blankets, handkerchiefs, and the like, which were obtained from the smallpox hospital maintained near Fort Pitt. Amherst wrote in a letter dated July 16, 1763, "You will do well to try to innoculate [*sic*] the Indians by means of blankets, as well as to try every other method that can serve to extirpate [them]." An outbreak of smallpox, previously unknown among the Ohio and Shawanoe tribes, killed them in great numbers during late 1763 and early the next year, although it is not clear whether the outbreak was caused by the infected gifts or coincidentally by some other contact with the white settlers.

Variability and Prediction

If the factors influencing outbreaks are so well known, why can't we do a better job of predicting epidemics from existing probability models? To put it statistically, there are too many parameters and they vary too much, so a particular prediction has too much uncertainty to it. Scientists may know a lot about a microorganism because it has been studied for a long time. They may know the molecular structure of its surface coating, the chemical structure of its toxins, a great deal about its metabolism, even its DNA sequence—but who knows the probability that an infected might meet a susceptible, whether in the Euphrates Valley or the New York City subway system? And what about the probability that the susceptible gets the disease, a prerequisite for transmitting it to others? This depends in part on the infectivity of the germ, which depends, in turn, on various chemical and structural details governed by DNA and thus subject to variability. How many germs are sneezed out? How many breathed in? What is the minimum infectious dose? This latter parameter is estimated for many germs by those who study the possibilities of germ warfare. But in natural settings, variability from person to person (and even within one person under various circumstances) comes into play, and there is simply too much uncertainty because of the inherent variability in so numerous a group of parameters.

For diseases with vectors—other animals that carry the disease to humans—the steps needed for the chain of causation may have been identified and scientifically demonstrated beyond the shadow of a doubt, as in the case of plague. The life cycles of the organisms involved are known in detail. Yet the uncertainties around our estimates of each parameter are so great, and the likelihood of the estimate being accurate is so small, that prediction is impossible. In overwhelming situations, such as when enormous numbers of plague victims and rats are traveling around Europe, or free blankets with smallpox are being distributed to a "virgin" population, you don't need a lot of information to realize that a problem is coming. But the situation is rarely so straightforward, and when it is you usually don't know it until after the fact (and even then only in broad outline). In fact, in epidemiology there are far more numerous factors, and much less understood about them quantitatively, than is the case in nuclear physics. And that's why we can successfully build an atomic bomb based on calculations of chain reactions yet cannot predict epidemics.

Profiting from Predictable Probabilities

Oddly enough, some sequences of events are governed very little by biological or physical parameters that can be known and estimated but, instead, are governed almost exclusively by what we might call random chance; and yet in some ways they are among the most precisely predictable phenomena of all. Think of gambling, for example. No team of scientists is needed to study the physical properties of coins, cards, or roulette wheels, yet it is possible for a casino or a lottery to predict in advance quite accurately the numbers of winners and the amounts of the payoffs in the long run. No need for lots of detailed scientific knowledge here—just statistical regularities, because there are few individual phenomena whose variability will have major impacts on the outcomes. The roulette wheel is so much simpler than many other aspects of life.

The sequence of events on a roulette wheel differs from the sequence of events in an epidemic in another important way. Each spin, each outcome, is independent of the one before. The roulette wheel has no memory. If the wheel is not "fixed" in order to cheat—if it is a fair wheel—then whether you win or not on a certain turn has no influence on your next turn. How different this is from an epidemic model, in which it certainly is relevant whether the person sneezing on you got infected from the person who last sneezed on him or her!

The roulette wheel was devised by Blaise Pascal, the eighteenth-century French philosopher-scientist, who was investigating the concept of perpetual motion. The device he came up with has meant fortune or ruin to thousands of people. The fortunate ones have runs of good luck; the unfortunate do not. If there is no memory in a roulette wheel and the spins are independent, how come there seem to be "runs" of good luck? If we use the laws of probability, can we predict such sequences of events? To answer these questions, let's look at one "run" of good luck in detail. To illustrate how statistical probability alone can cause runs of this sort, pick an example that we can model as a simple, purely random process. A run of good luck for a team playing some sport such as soccer or baseball wouldn't be a good example for this purpose since there are too many complicating factors operating at once, such as the teams that happen to be arrayed against them. Even the timing of the game relative to other games, or the enthusiasm of the fans, might have a bearing on a player's performance; and the teammates all must interact, adding more complication to the model. All of this occurs *on top of* the inherent random variability in performance that we might call pure luck. So let's stick to a game of chance to keep it as simple as possible and see if we can expect or predict runs of good luck.

The roulette wheel at a casino in Monte Carlo has 37 slots numbered 0 through 36 into which the little metal ball will fall at the end of a spin. Bets can be placed on odds or evens. Of course, if we only consider 1 through 36, there are 18 odd and 18 even numbers, so we would expect the house to break even: on average, half the bets would be placed on odds, half on evens, so the house would collect from half the bettors and pay out to the other half. But the presence of the 0 is the means by which the house makes a profit on roulette. Zero is considered neither odd nor even; if the ball lands on 0, the house collects all the amounts wagered. Thus, 1 of the 37 slots affords the possibility that the house wins. The ball has a 1/37 probability of landing on 0, and the wheel is spun roughly 500 times a day, so, on average, 13 or 14 times per day the entire game's bets accrue to the house. Roulette wheels in American gaming houses favor the house even more strongly: there are both a 0 and a double 0, so the chance is 2/38 that the number chosen is "neutral" and the house collects all wagered amounts. In either case, the advantage in favor of the house is pretty impressive (1/37 is 2.7%; 2/38 is 5.26%). And these percentages can be counted on as quite reliable predictions in the long run. But once in a while, a sequence of events may occur that shocks bettors and house management alike.

A case in point occurred in 1873, when an English mill engineer by the name of Joseph Jaggers won a huge amount of money at the beaux-arts casino in Monte Carlo. His assistants went to the casino the day before he did and noted down all the numbers that came up during that day. Jaggers looked over the numbers, searching for evidence of non-random patterns. Five of the six roulette wheels in operation were perfectly normal. The sixth, however, had nine numbers that came up far more often than chance alone would suggest. The next day, Jaggers came to the casino and played those nine numbers. By the end of the fourth day, he was sitting on about $300,000. The year Jaggers died, the English music hall performer Charles Coburn had tremendous success with the song "The Man Who Broke the Bank at Monte Carlo."

Jaggers's good fortune was not really a triumph of mathematics, but of physics. A small scratch in the roulette wheel caused those nine numbers to come up more frequently than, statistically speaking, they should have. Since then, the roulette wheels in the casino at Monte Carlo have been examined daily by engineers to ensure that all numbers come up equally frequently.

The Gambler's Fallacy

One incident at the casino at Monte Carlo stands out as the most extreme outcome ever seen there, and this one *wasn't* due to a physical flaw in the wheel: on August 18, 1913, evens came up 26 times in a row. Given that on a single spin there are 18 evens out of 37 possible outcomes, the chance of a single even outcome is 18/37, or about 0.486 (same as for odds). The probability of 26 in a row is $18/37 \times 18/37 \times 18/37 \ldots$, 26 times, which is 0.000000007, or roughly 1 in 142,857,000. What a run of luck—or rather, it would have been, had anyone been "stout of heart" enough to stay the course through all 26 evens, and prescient enough to quit right on time. However, in the expectation that an odd number was somehow overdue, there was a huge stampede of bettors abandoning evens at various points, until no one was left to profit from the run except the house.

Was it luck? Did Destiny provide a good opportunity for a favored gambler, who ignorantly spurned her offer? Remember that roughly 500 turns of the wheel occur daily, and 4 or 5 or more may be turning at once. The casino is open almost every day and has been there for more than 125 years. Eventually, you would see a run like that, or indeed of any specified size. Runs of "good luck" can be generated by no special mechanism at all; no explanation is necessary beyond the laws of chance and the multiplication of those successive probabilities. It wouldn't be your best guess as to what would happen next, because it *is* exceedingly rare, but a one-in-a-million event, even a one-in-five-hundred-million event would be expected to occur in a large enough sample of outcomes. To this extent it is predictable—we can estimate the probabilities and, given a large enough sample size, see extreme series of events in accord with mathematical expectations. It is easier to estimate the probability of some strange-seeming event in the case of the roulette wheel than it is to do so for an outbreak of a rare disease, or the outcome of a baseball season. It's hard to imagine a simpler chain of probabilities than $18/37 \times 18/37 \ldots$ over and over again.

Note that we can do this calculation in this way precisely because with a fair wheel the events (outcomes on each spin) are independent; we need to know little because the probabilities are unchanging, since no event is contingent on a prior event. So a single probability will do for our model. But another implication of this independence is a source of despair to the gambler: since one outcome has no bearing on the next, it gives the gambler no information on what will happen next. Beliefs such as "We are overdue for odds" are known as the gambler's fallacy. In roulette, every time the ball begins its travels anew, it's a fresh start, and the fresh start is the same on the first spin as it is on the millionth. It's a bit of a paradox: predictions can be made from probability theory that tell us, rightly, to expect some otherwise surprising runs in the behavior of that little ball, when the outcomes are considered en masse; yet at the individual level of detail, actual predictions are impossible. These are statistical predictions, only good for forecasting patterns in the group of outcomes.

Distributions

The forecasting of the patterns of independent outcomes can be made much easier if a *probability distribution* is available that serves as a model of the events under study. A distribution is a handy tool. We can consider it as a list of possible outcomes, together with the probability of each of those outcomes. An everyday example of a probability distribution is the set of possible outcomes when throwing two or-

Table 1.1. Distribution of possible outcomes when throwing two dice

	Sum	Ways to Get That Sum (first die, second die)	Probability You Get That Sum
	2	(1,1)	1/36 or 0.028
	3	(1,2);(2,1)	2/36 or 0.056
	4	(1,3);(3,1);(2,2)	3/36 or 0.083
	5	(1,4);(4,1);(2,3);(3,2)	4/36 or 0.111
	6	(1,5);(5,1);(2,4);(4,2);(3,3)	5/36 or 0.139
	7	(1,6);(6,1);(2,5);(5,2);(3,4);(4,3)	6/36 or 0.167
	8	(2,6);(6,2);(3,5);(5,3);(4,4)	5/36 or 0.139
	9	(3,6);(6,3);(4,5);(5,4)	4/36 or 0.111
	10	(4,6);(6,4);(5,5)	3/36 or 0.083
	11	(5,6);(6,5)	2/36 or 0.056
	12	(6,6)	1/36 or 0.028
Total			1

dinary dice. Since there are 6 possibilities on one die and 6 on the other, there are a total of 6 × 6, or 36, possible obtainable combinations, and all 36 combinations are each equally likely if the dice are fair and independent. However, the various combinations add up to certain sums ranging from 2 to 12, and the sums are considered the outcomes when we toss dice. These sums are not equally likely because certain sums can be arrived at in more than one way. Table 1.1 shows the outcomes, how you can get them, and the probability of each.

Many probability distributions can become too complicated to set up in the form of a table, because there are too many possible situations. One such distribution is the binomial distribution, so called because it gives the probabilities of outcomes that must occur as one of two categories. However, in the simplest situation, such as coin tosses with equal likelihood of heads or tails, values for the distribution can still be calculated using a tabular format that was developed by Pascal, the inventor of the roulette wheel. This format, known as Pascal's triangle, appears in table 1.2.

Table 1.2. Pascal's triangle

Number of Trials	Numerator for Probabilities	Denominator for Probabilities
1	1 1	2
2	1 2 1	4
3	1 3 3 1	8
4	1 4 6 4 1	16
5	1 5 10 10 5 1	32
6	1 6 15 20 15 6 1	64
7	1 7 21 35 35 21 7 1	128
8	1 8 28 56 70 56 28 8 1	256

Here's how to read the table: Suppose that the number of tosses, or trials, is 1. On the corresponding line, the two 1s in the triangle provide the numerators for the two possible outcomes, heads or tails. The total number of possible outcomes is, of course, 2, so the denominator for calculation will always be 2, and the probability in both cases is 1/2. With two trials, as shown on the next line, there are three possible outcomes. Two are represented by 1s and occur with a probability of 1/4; these are the extremes, "two heads" and "two tails." The third outcome is "one head and one tail." There are two ways to get such an outcome (two different orders in which they can come up), so that outcome is represented by a 2 and has the probability of 2/4, or 0.5. Finally, let's suppose that we are going to toss a coin four times, so we have four "trials" or experiments. There are two possible outcomes, heads or tails, on the first try, and two apiece on the second, third, and fourth tries; thus, the total number of possible combinations we might see is $2 \times 2 \times 2 \times 2$, or 16. Therefore, 16 is the denominator. The triangle provides the numerators and lets us calculate probabilities for all 16 combinations: 1/16 that we get all heads; 4/16 that we get 1 head, 3 tails; 6/16 that we get 2 heads, 2 tails; 4/16 that we get 3 tails, 1 head; and 1/16 that we get all tails. Note that since the triangle is symmetrical, it doesn't matter which side you call heads or tails.

In the days before calculators, it was extremely useful to be able to construct a table like this, and it required little memory to set it up. Once you write 1s for the outer borders of the triangle, any other number needed can be obtained by adding the two numbers closest on the diagonals above it. The process can be repeated for any arbitrarily large number of trials.

But what happens when it's not a 50-50 split? As great an invention as Pascal's triangle was, it only works for "even splits," and there is such a huge variety of probabilities (and numbers of trials) that it would not be practical to generate the enormous books of tables that would be required to cover every possible situation. Here is an example for which we can really appreciate the economy inherent in the mathematical expression known as an equation. From one little equation, we can generate all possible binomial distributions, for any desired number of trials and for any underlying probability of event. Suppose that a pharmaceutical company claimed that a new drug, taken as directed, results in an 80% therapeutic success rate. A physician has three separate and independent patients, and they use the new medicine as directed, but only one patient gets better. Is this surprising? Is the observed success rate of 33% evidence against the company's claim? We can use the binomial formula to answer such questions.

The Binomial Formula

Suppose that p is the assumed underlying probability of a successful treatment; we are told by the company that $p = 0.8$. Then the probability of a treatment failure is called q, and $q = 1 - p = 0.2$. The number of trials is called n, and $n = 3$ in this example; the number of outcomes that are therapeutic successes is called r, and $r = 1$. The question is, How likely is it to see 1 of 3 therapeutic successes when the underlying true rate is 0.8? The binomial formula is

$$P(r) = \frac{n!}{r!(n-r)!} p^r q^{(n-r)}$$

so for our example,

$$P(1) = \frac{3!}{1!(2!)} 0.8^1 0.2^2$$

The exclamation point does not mean that "n" is to be shouted emphatically. It is read as "factorial" and represents the sequential multiplication of all integers from a given number down to 1; thus, $3! = 3 \times 2 \times 1 = 6$. For the record, 0! and 1! are equal to 1. After a little arithmetic, it turns out that $P(1) = 0.096$. Thus, even with an underlying 80% therapeutic success rate, in any given set of three patients, you would have nearly a 10% chance of seeing just a single success. The intuitive expectation that two or three should be therapeutic successes would indeed be your best guess as to the outcome, since $P(3) = 0.512$ and $P(2) = 0.384$; but having a distribution that matches the situation allows the precise calculation of the probability of every possible outcome. Knowledge of these probabilities is the basis of quantitative predictions before an event occurs, as well as a basis for judging the likelihood or rarity of events once they are observed.

Why does this formula work, and why didn't we need it when we calculated the probability of getting 26 in a row on Monte Carlo's roulette wheel? First of all, we *were* actually using the binomial distribution then, but we didn't need the formula to figure it out. Here's why: the fraction part would be 26!/(26!0!), since there are no odd outcomes and $(n-r) = 0$. But $0! = 1$, and $26!/26! = 1$. The fraction part will always equal 1 when the outcomes are uniformly one way or the other, so the fraction part isn't needed for the calculation in such a situation.

The fraction part tells you the number of combinations that can provide a particular outcome. For uniform outcomes, such as all evens, there's only one way: each and every spin must come up even. By contrast, there are 26 ways to get a single odd spin out of 26 trials: the odd one can be first, second, third, . . . all the way up to 26th. When calculating the probability of a single spin being odd and 25 being even, the fraction part would be $26!/(25!)(1!)$ and would reduce to 26.

Similarly, while we *did* use $(18/37)^{26}$, the q's superscript $(n - r)$ is 0, once again because there are no odd outcomes whose likelihood has to be factored in. Thus, calculating the probability of a long string of evens is easy to do, since you just take the probability of a single even outcome "times itself" as many times as there are evens; however, many people do not realize that that is just an easy, special case of the more general and somewhat more complicated binomial distribution.

The binomial model provides a model of the behavior of the roulette wheel that is an accurate representation of the outcomes (odds versus evens) that you would actually observe in a long series of spins. The binomial is not quite an accurate representation of what is observed with a long series of coin tosses, because most of the world's coinage is slightly unbalanced—a little more than half the weight may be on one side or the other, owing to the slightly different volumes taken up by the design on each side. For example, the English statistician J. E. Kerrich was trapped as a prisoner of war during the German occupation of Denmark during World War II. Having plenty of time on his hands, he tossed a coin 10,000 times and got 5,067 heads—in other words, 50.67% instead of 50%. Realistically, we expect only small deviations from 50-50 and, moreover, the coin tosses are still independent. Thus, the coin-tossing gambler who feels an imminent reversal of fortune is "overdue" after a long run is still mistaken. However, the binomial model, especially with a 50-50 split assumed, is not an adequate model for predicting the frequencies of boys and girls in a series of births.

First, demographers know that the sex ratio at birth ranges from 104 to 107 boy babies per 100 female babies, and that this finding is quite general across societies and time periods. Occasionally, the sex ratio is reportedly 110 or even higher, as in some regions in rural China. Some observers believe that the elevated ratio there reflects infanticide originating in a strong preference for boys, coupled with a forceful governmental population control policy providing sanctions for families with more than one child. Others think that the governmental policy provides the motivation, but that a failure to report and register the births of females in rural areas results in the official statistics (but not the population) being skewed toward boys. China aside, it is a biological fact that a typical proportion of boy babies is 105/(105 + 100), or 51.2%; a typical proportion of girl babies is 100/(105 + 100), or 48.8%, and the difference between the two is 2.4%. Thus, if you always bet on boys being born, and an opponent always bets on girls, your profit margin in the long run would be about the same as that garnered by the Monte Carlo casino on a roulette wheel.

However, boys have a much higher mortality rate at every age, from birth to 100 and beyond, so the surplus proportion of boys diminishes with every passing year of age, until it becomes reversed and widows far outnumber widowers. Indeed, the balance between births and excess mortality so impressed the statistician Johann Peter Sussmilch that he published a book in 1741, whose title translated into English is *The Divine Order as Derived from Demography.* The surplus of boys at birth was, he said, evidence of the wisdom of the Creator "thus compensating for the higher male losses due to the recklessness of boys, to exhaustion and to dangerous tasks, to war, to sailing, and to emigration. . . . [He] also maintains the balance between the two sexes so that everyone can find a spouse at the appropriate time for marriage." Not everyone did get married, not even in Sussmilch's day, and we might ask the mortality differential to end at marriage, so that the years lived in widowhood might be less on average. These aspects of the divine plan are not addressed in Sussmilch's book.

The preceding quote by Sussmilch is from the translation provided by my friend and colleague Anouch Chahnazarian, published in *Social Biology* in 1988 shortly before her own untimely death a few years after completing her Ph.D. Her article provides an extensive examination of factors affecting the sex ratio at birth. This includes maternal levels of the hormone gonadotrophin: these levels are inherited and may vary by family and even by ethnic group. For example, she states that persons of African ancestry "having higher levels of gonadotrophin, would have a higher probability of having girls, and hence a lower sex ratio at birth," than persons of European ancestry. This expectation is validated by the findings of repeated studies of millions of births, in which the sex ratio of African-American populations at birth ranges from 102 to 104. In addition, births conceived early or late in the menstrual cycle tend to have a higher sex ratio, earlier births within a family tend to be males, and births to younger fathers tend to be males. Hence, a "coin toss model" is wrong for births, but not only because the chances deviate from 50-50: the binomial is also a poor model because successive births from one woman are not independent. Her characteristics, hormonal or ethnic, for example, are more similar from one pregnancy to the next than would be the case for two randomly selected pregnancies from two different women.

In any case, a simple distribution like the binomial sometimes cannot be used to provide reliable estimates of expected outcomes for a different kind of reason: in some situations, many factors influence the outcome, and each one has an unknown distribution. Then a good statistical model would require knowing much more than a single probability of success and the number of trials. In such situations, modeling may be limited to a crude and imprecise model with rather imperfect prediction. However, an imperfect prediction is often better than none and may have much practical value.

This situation is especially common when human behavior, rather than human physiology or medicine, is the subject of statistical prediction. For example, people in the business of marketing know that potential customers grouped by zip code and some other demographic data have an increased or decreased propensity to buy a certain item compared with other groups. Not every potential customer will "bite," but the predictions are good enough to increase the yield of a marketing campaign and cut down on the costs of expensive catalogs or phone calls, or advertisements futilely directed at those who are unlikely to buy. Once again, you can predict the group outcomes in terms of proportions buying (that is, the statistical probability or rate of buying), but not an individual's decision. No marketers worth their salt would propose to predict an *individual's* behavior, but they can predict a large group's behavior, and their profitability may depend on it. So while you may think of unsolicited mailings as junk mail, in a successful campaign these mailings have been very carefully chosen to be sent to *you.*

I can think of two professions that do have the aim of predicting of the behaviors of specific individuals: forensic psychiatry and astrology. How well founded are their claims of making useful predictions? Remember that all practitioners of medicine, not only forensic psychiatrists, are constantly asked to make predictions. Patients constantly ask them what to expect, and choice of treatment is driven by an implicit prediction of what will happen under various scenarios. When the treatment is new and the disease or illness is poorly understood, there is more uncertainty and the prediction is of dubious accuracy. For example, in certain cancers that are currently incurable by standard treatments, a clinical trial of a new medicine may be conducted in order to determine whether that medicine lengthens life, or makes it more pleasant. There must be some reason for a prior expectation that it will, or the experiment wouldn't be conducted; on the other hand, there must be great uncertainty, or it wouldn't be necessary to conduct an experiment. In other situations, the prediction is excellent. Someone with a strep throat treated with antibiotics will almost surely get better. But all predictions have some degree of uncertainty. Even with a strep throat, occasionally the patient may have come to the doctor too late, have a virulent, antibiotic-resistant infection, or suffer from an immune disorder resulting in an overwhelming susceptibility to the strep bacteria. Thus, we often hear doctors say, "I think you'll be okay," but that's a general prediction or a best estimate. They would never say your recovery is an absolute *certainty*, if pressed. And when not just the physical well-being but the behaviors of an individual are involved, as in psychiatry, one might expect prediction to be most difficult of all.

Predicting Criminal Recidivism

Forensic psychiatry is that branch of medicine that deals with psychiatry and the law. Forensic psychiatrists are often asked to decide whether it would be dangerous to release a particular person who is being held in a prison or mental hospital. These psychiatrists are being asked for predictions about whether socially undesirable acts—so severe as to have caused incarceration—are likely to be repeated if the person is returned to society at large. Naturally, statistics concerning the incidence of violence among people with various psychiatric diagnoses will come into play here; for example, patients suffering from paranoid schizophrenia are generally more dangerous to others than are patients suffering from depression. But an individualized prediction is much more valuable than such generalities, and it must take into account the multifaceted aspects of a particular personality, the history of the course of development of that mental illness in that person, and the specific acts committed and their circumstances. There is likely to be much variability from person to person in these factors, which will have a significant impact on the decision. Indeed, the decision is so fraught with ambiguities that it is common for respected forensic psychiatrists to argue forcefully on opposite sides in a courtroom, which has led to the formulation of Holland's law of testimony: For every psychiatric expert there is an equal and opposite expert. By contrast, the decision is pretty straightforward once the diagnosis is established; the sensitivity of the bacteria to various antibiotics can even be determined in advance. The decision has comparatively little ambiguity because the normal variability from person to person is not too likely to have a significant impact on the outcome of treatment.

So how well can future behavior be predicted by forensic psychiatrists? Two studies provide some clue. The first, published in the *American Journal of Psychiatry* in 1972, concerns patients who had been held in the New York State Hospital for Insane Criminals, in Dannemora. One of the prisoners, Johnnie K. Baxstrom, sued to be released, saying that the explicit purpose of his incarceration was for treatment and that he was being subjected, unconstitutionally, to incarceration without the mandated treatment. In addition, his rights were violated because "the administrative decision to retain Baxstrom in Dannemora was made before any hearing was afforded to Baxstrom and was made despite the otherwise unanimous conclusion by testifying psychiatrists . . . that there was no reason why Baxstrom could not be transferred to a civil institution," according to the U.S. Supreme Court decision of 1966 (*Baxstrom v. Herold*). Eventually, Baxstrom and 969 other patients were released. Although previously remanded to incarceration at Dannemora because of predictions of imminent dangerousness, all 970 were subsequently treated at a regular psychiatric hospital.

Four years after the transfer to a regular hospital, half had been released and gone to live in ordinary communities; the other half remained hospitalized or were rehospitalized later. More than three-fourths of the 970 had gone four years without committing any assault whatsoever against any hospital staff member or the general public; the remainder had committed some infraction (although the majority of these infractions were considered minor from the medicolegal standpoint). The question is, Does the evidence, provided by this unexpected release for legal reasons, point to good prediction in the case of the 970 sent to Dannemora? Half survived successfully in the general community, the other half in an ordinary, rather than criminally oriented, psychiatric institution. More than three-fourths showed no further evidence of dangerousness, despite the original psychiatrists' judgment that overriding concerns for public safety justified involuntary incarceration of each of these patients. Some people conclude from these findings that proper treatment in ordinary hospital settings or in communities is all that may be needed for adequate control of some violent mental patients who have been arrested (and similar figures have been found in similar circumstances upon release from other U.S. mental hospitals for criminals). On the one hand, it seems that more than three-fourths of the Dannemora patients were deprived of their liberty and incarcerated unnecessarily, because the prediction of dangerousness was *not* warranted for *them*; on the other hand, *as a group*, they *are* dangerous, having a higher rate of committing some further assault than the general population. The dilemma that this situation presents for public policy is due in part to the statistical property of having predictive power for a group but not for an individual.

In 1972, H. L. Kozol and colleagues published another study of the predictive power of physicians' judgments of dangerousness. This time, the outcome under investigation was the rate of repeat offenses committed by sex offenders who had been detained for such crimes and later released. One group was released by the courts in accordance with doctors' judgments that it was safe to do so; the group's rate of repeat offenses was 6.1%. Another group was released by legal tribunals, *against* doctors' judgments, and the rate of recidivism was 34.7%. Therefore, it seems that the doctors were much better than the legal bodies at predicting dangerousness, but they weren't perfect either—they were wrong in about 6% of cases. Although a small percentage, this is an important error in prediction, of course, since these 6% went on to assault someone again. Is the use of doctors' predictions resulting in release decisions that are adequate for society? Setting aside issues of punishment, should all of the sex offenders have been permanently incarcerated, so that the detention of 94% would serve the purpose of preventing the recidivism of the other 6%? Would this be justice? Or is it acceptable to have a process that allows offering freedom to not just the 94% but also to those 6% who go on to repeat their heinous crimes?

The public is well aware that certain convicted violent criminals, specifically sex offenders, may be released by the legal system into the community, and that such offenders may have a fairly high likelihood of repeating their crimes. From this awareness arose a political movement that successfully sought the passage of laws requiring the names of convicted sex offenders to be listed in a publicly available registry, once the offender is no longer incarcerated. The movement gained momentum from the outrage over circumstances surrounding the brutal rape and murder of 7-year-old Megan Kanka of New Jersey. Her parents first learned that a neighbor was a twice-convicted sex offender when he was arrested for the deadly attack. Each state was subsequently required to meet U.S. federal guidelines by passing Sex Offender Registration Acts in the mid-1990s. In most locations, the new set of rules is informally called Megan's law, after the incident that served as the catalyst for legislative action.

The level of risk of further criminal offenses may vary from one released offender to another, and Megan's law dictates that the level of risk shall determine the amount of information released and the type of notification permissible. Recommendations about level of risk are made to the court as a required part of the process when a convict is being released. The recommendation is made based on the Sex Offender Registration Act Risk Assessment Instrument. The instrument, or questionnaire, categorizes risk into one of three levels, based on a point system, and if a convict is classified as a level 3 offender, information about the convict including name, age, address, criminal history, and photograph may be posted, even on the Internet. The point system used is very detailed. For example, unwanted touching through clothing equals 5 points; sexual intercourse equals 25 points. If the crime occurred less than 3 years ago, 10 points are added. If there were two victims, that's worth 20 points; three or more victims, 30 points. Four circumstances automatically garner a level 3 classification: prior felony conviction for a sexual crime, serious injury or death of the victim, threat of further criminal acts made by the criminal, or clinical judgment that the individual lacks self-control because of specific psychiatric abnormalities. This is an example of public policy being driven by the public's concern over the much-less-than-perfect predictive value of psychiatrists' judgments and the premature release of offenders. Yet, paradoxically, those agitating for Megan's law must have accepted that statistical predictions are of some value, for the compromise solution entails notification based in part on statistical categorization of risk.

There will always be unpredictable variability in individuals even with a seemingly similar criminal and psychiatric profile. There are also always differences in the contributing factors that, through a long chain of causation, end up in the commission of crimes. There will probably also always remain interrater variability among psychiatrists in their assessment of an individual's dangerousness. There can never be a perfect checklist with a scoring system allowing us to free safely all those with "dangerousness scores" lower than some cutoff value. This type of problem, therefore, cannot have a permanent, objectively correct, agreed-upon resolution, because statistical issues concerning prediction for groups must necessarily have an impact on medicolegal judgments.

Do Our Faults Lie in the Stars?

Astrology is another field that seeks to make useful predictions about human behavior, as does forensic psychiatry, and should be subjected to the same criteria of statistical proof. The low value of astrology in predicting specific, unambiguous life events or sequences, or even personality traits, has been noticed since ancient times. A case in point is the commentary of the Roman senator Cicero, who lived from 106 to 43 B.C.E. In his book *De Divinatione* (I'm using Falconer's translation), Cicero notes that "Plato's pupil, Eudoxus, whom the best scholars consider easily the first in astronomy, has left the following opinion in writing: 'No reliance whatever is to be placed in Chaldean [Babylonian] astrologers when they profess to forecast a man's future from the position of the stars on the day of his birth.'" Cicero's high opinion of Eudoxus (a Greek who lived around 400 B.C.E.) has been borne out over time. Eudoxus not only proved key theorems in geometry; he also produced the first Greek map of the stars and was the first Greek to calculate the motions of the planets using a geometrical model employing the mathematics of concentric spheres, an important calculational tool as well as a conceptual advance.

Cicero sets forth many reasons for doubting astrology. The mechanism cannot be true because, as he notes, at the same moment in time the stars will not be in the same positions for all observers on Earth, but rather vary according to where they are. He also notes that appearance, gestures, habits, and lifestyles of children resemble their parents. These traits, and those differences in body and mind-set "which distinguish the Indians from the Persians and the Ethiopians from the Syrians" are evidently "more affected by local environment than by the condition of the moon." More important to him than this general reasoning is the lack of predictive power of astrology: to him, "the fact that people who were born at the very same instant are unlike in character, career, and in destiny, makes it very clear that the time of birth has nothing to do in determining the course of life." Most tellingly, he asks, "Did all the Romans who fell at [the battle of] Cannae have the same horoscope? Yet all had one and the same end." Cannae was a truly crushing defeat for the Roman army. In 216 B.C.E., Hannibal, leading the Carthaginian troops of about 50,000 men, destroyed the much larger Roman army, killing more than 60,000 of its 80,000 soldiers. Cicero has a point: if horoscopes could not be used to predict death at Cannae, they could not be said to possess predictive value.

This is implicitly a statistical argument. A Roman's chance of dying at Cannae is independent of when he was born (the rates are the same for all birth dates and times); and, conversely, among those soldiers who died in battle there, no preponderance (no elevated rate or probability) of a particular horoscope could be seen. These days, the evidence against astrology is more explicitly statistical, and the information is better documented and better controlled. A study published in 1985 in the British journal *Nature* illustrates this point and demonstrates the strength of the statistical evidence against astrology.

Unlike Cicero, the research team conducting this modern study consulted professional astrologers of high standing in that field, in order to avoid testing "the scientists' concept of astrology" rather than "astrology as practiced by the 'reputable' astrological community." These astrologers were selected from a list provided by the National Council for Geocosmic Research, an organization respected by astrologers worldwide. Twenty-eight astrologers agreed to participate and to produce natal charts showing positions of heavenly bodies at birth, together with the resultant personality descriptions, for more than 100 college-educated volunteer subjects from the general public. Despite Cicero, astrologers do know that the stars are different from one location to another at the same point in time; they were therefore provided, as requested, with exact location, as well as date and time of birth. This information was included in the study only if documented by birth certificate, hospital records, or other records contemporary with the birth.

The study had two parts. First, each birth record resulted in a natal chart and a written interpretation by an astrologer, who described the personality traits that the corresponding individual subject was supposed to possess. Next, the names were removed from these interpretive documents. Each volunteer was then presented with three personality descriptions: his or her own as well as two others chosen at random from the remaining lot. The task for each subject was then to pick the description of himself or herself. By chance alone, one would expect 1/3 of their choices to be correct. This was the scientists' expectation. The astrologers felt that their work would result in an identification rate better than 50%.

These predicted rates of matching represent estimates. No one said that rates had to be *exactly* 1/3 or 1/2 to prove their point, of course.

Sampling fluctuation is to be considered, and this is a phenomenon produced by random chance. For example, even if a coin produced a 50-50 split of heads and tails in the long run, in 10 tosses one might well have exactly 5 heads and 5 tails, but it would not be surprising to find 6 of one and 4 of the other. Indeed, *always* finding exactly 5 heads and 5 tails in each set of 10 would be shocking. The point is that in a single experiment your expectations are centered on a certain number; the closer the actual outcome is to that number, the more the finding is supportive of the viewpoint that had resulted in choosing that number.

In the case of the test of astrology's success, complete data from both the test subject and the astrologer were available in 83 cases. How often did a subject recognize the personality description resulting from his or her own natal information? Correctly matching choices were made by 28 of the 83 subjects, or about one-third of the time (33.7%, in fact). Thus, it would appear nothing special is going on beyond chance matching. Perhaps astrology did not produce recognizable descriptions of subjects from their natal information. However, this is a weak measure of astrology's success. Astrologers or even scientists might well argue that the outcome merely demonstrates that people are not good at picking out descriptions of themselves.

The second, stronger part of the study involved the examination of a different type of outcome. Individual subjects each filled out the California Psychological Inventory (CPI), a bank of more than 400 questions about various preferences, situations, and traits that has been around since 1958. The questions are in the form of statements, with respondents labeling each statement as true or false. H. G. Gough published some examples of typical questions in *Psychological Reports* in 1994. These included

- "If the pay was right I would like to travel with a circus or carnival."
- "I would never play cards (poker) with a stranger."
- "Before I do something I try to consider how my friends will react to it."

The CPI produces numerical ratings on a number of subscales such as dominance/passivity, self-control, tolerance, and flexibility. This test does not directly ask for characterization of one's own personality, and the results have been subject to external validation by comparison with psychologists' judgments concerning personality traits. Thus, the test is well respected by psychologists; moreover, it was selected from among the available personality traits specifically because the astrologers recruited for the *Nature* study agreed that the personality traits measured were closest to "attributes discernable by astrology." The test results are presented as a CPI profile that gives numerical results for the subscales.

In a mirror image of the first part of the study, the astrologers were presented with sets consisting of one natal information chart and three CPI profiles. One of these profiles came from the individual whose natal chart was provided. The two others were selected at random from the remaining lot. The task for the astrologers was to take the birth information and decide which of these three CPI profiles was the matching profile from the same individual. Once again the scientists' prediction was that the astrologers' rate of success at matching would be consistent with the 1/3 probability expected by chance alone. The astrologers, for their part, felt that astrology would be proven useful if they achieved 50% correct matches. There were 116 natal charts available to be matched with the CPIs. Thus, the scientists estimated that there would be 39 correct matches, while the astrologers expected to make 58 matches or better. The number of correct CPIs chosen was 40. In addition, astrologers were also asked to rate how confident they felt about each match, on a scale of 1 to 10. There was a strong tendency for answers to cluster around 8, but it was no different for correct matches versus incorrect ones.

Occasionally, people I meet socially will ask me my astrological "sun sign." Years ago, I used to tell them—a revelation invariably followed by a flash of "understanding" and a comment like, "Of course! That's why you have such-and-such a personality trait!" I always thought it odd that they had to ask, if it was so obvious. Why not just come up to me and say, "You must be a Leo," or "You're a Pisces, aren't you?" So I no longer reply with the sign, but rather a little challenge, because I am curious to see the outcome. When asked my sign, I reply, "*You* tell *me*." I've kept track of the answers, and so far all the signs are suggested to me in equal numbers. This is called an *even distribution* of outcomes, or sometimes a *flat distribution*, because in a bar chart the bars representing the signs would all be about the same height and no one sign would predominate. I've often thought about taking it one step further and giving an incorrect sign and seeing whether my fraud would be detected. So far, I've resisted the temptation.

My own experience of astrological predictions of sun signs, as well as the astrologers' inability to predict CPI outcomes from birth records, implies something about the distribution of personal traits across astrological categories. Whether the categories are broad, like the choice of sun sign categories, or narrow, like the multitude of possible natal information configurations, personal traits are more or less equally distributed across the categories. If you used the height of a bar graph to show the proportion of people with a given trait, and used a separate bar for each astrological category, the bar would be about the same height for each category. Similarly, the bars would be of about equal height if Cicero had made a graph showing death rates at the battle of Cannae, as in comparing sun signs or astrological configurations. When there is a flat or "even" distribution of data among categories, an individual's membership in a given category gives no useful information about the likelihood of a particular outcome for that person.

At the other extreme, some distributions of data provide lots of information about the likelihood of given outcomes. We've already become familiar with the binomial distribution and its use in estimating probabilities, but many situations involve continuous data. For example, blood sugar levels have a continuous distribution, but the distribution is shifted to the high side in the diabetic patient. Elevated pressure within the eyeball is a good predictor that the degenerative disease known as glaucoma is under way, and the higher the intraocular pressure, the worse it is. Many measurements like these in clinical medicine, and many in social science, follow a bell-shaped curve called the *normal distribution*. When measurements do follow a normal distribution, calculation of the probabilities of various measurements within groups is made easier, and prediction is improved. Applications of the normal distribution are the main subject of the next chapter.

2

Surely *Something*'s Wrong with You

A great performance form the Yankees' Derek Jeter helped crush the Baltimore Orioles at the stadium yesterday.

As I typed the previous sentence, my word processor automatically looked for errors in grammar and spelling and underlined the word *Jeter*. The ballplayer's correctly spelled last name was classified as a spelling error. On the other hand, there's a typographical error that was missed: the word *from* should have appeared, rather than *form*. Any system that classifies things into two categories can make two kinds of mistakes, called "false positives" and "false negatives." Underlining *Jeter* is a false positive—his name shouldn't be flagged as a problem, yet it is. Failure to flag the word *form* is an example of a false negative—it should have been noticed as a problem, yet it was not. The prevalence of this type of error in the output of word-processing programs has ensured the continued employment of copy editors, because (at least for the present) people can recognize why *Jeter* was right and *form* was wrong much more reliably than computers can.

The inevitable phenomenon of false positives and false negatives can have important consequences. For instance, they may pose a serious threat to proper diagnosis and treatment of illness. Blood tests provide a perfect example. In days gone by, a visit to the doctor meant that leeches would be applied, or a bleeding performed, to rid the body of impurities. Nowadays, a tube of blood is drawn and sent to a clinical laboratory for analysis. A few hours or days later, the results come back, and perhaps some of the lab values will be flagged as abnormal, meaning that they are outside the normal range. Your doctor may remark, "Something's not right here: your blood levels of cholesterol and calcium are high." These elevated serum levels may be indicative of an increased risk of heart disease and of metabolic disorders, respectively. But do abnormal findings on laboratory tests necessarily mean that something is wrong with you? Or could the lab's results be false positives or false negatives? If so, how prevalent are errors?

To start with, we need to know what "normal" means in this context, so we need to know how normal ranges are set in medicine, and how people get flagged as abnormal. Here's a typical process: A large group of presumably disease-free individuals have blood drawn, and the distribution of values for such variables as cholesterol or calcium is established. (In chapter 1 we defined a *distribution* as a set of possible outcomes together with the probability of each of those outcomes.) With the distribution of lab values in hand, the next step is to decide what constitutes an extremely high or low value. Usually the 95% of values in the center are considered unremarkable. The 5% of values that are most extreme are used to set cutoffs for the normal range (that is, the $2\frac{1}{2}$% highest plus the $2\frac{1}{2}$% lowest observations).

Although some normal people apparently have such extreme values, lab values way off center are worth further investigation, because they are probably more typical of other distributions, such as those of people with disease—they are certainly not typical of normal individuals.

This common procedure is not based on any detailed knowledge of disease processes. It's not as if scientists are able to determine the distribution of lab values that should be seen in normal individuals or in case of disease based on the pathology involved and nothing else. It's all based on typical observed population values. Indeed, 5% of absolutely normal people will inevitably have lab values classified as "extreme" by this method. What's worse, the more testing that's done, the more likely a mistaken classification will occur, a person will be categorized as "abnormal" on some variable, in the absence of disease. False positives on medical tests are often caused by this phenomenon, especially when a large battery of tests is ordered for a particular patient.

We can use the binomial distribution to quantify the chances of false positives here. Suppose that 18 lab tests are conducted on a disease-free individual, and there's a 95% chance of being classified as normal by any given test. The chance that all 18 will be normal is 0.95^{18}, since we assume that fluctuations in the tests are independent in normal individuals. The combined probability of all outcomes is, of course, 1.0; the combined probability of all outcomes *excluding* the uniformly normal findings on all 18 tests is thus given by $1 - (0.95^{18})$. Hence, the chance that at least one lab test will show an abnormal reading turns out to be greater than 60%. So, if you go to a doctor and have a lot of tests done, it is likely that some of them will indicate a suspicious finding and that further tests will be required.

This process does not get repeated indefinitely, however, because the odds are small that successive independent tests for a given condition will all be positive, unless, of course, something is truly going wrong. In addition, second-round tests are usually better at discriminating between those with a given illness and those without it. Usually such tests are more expensive, and that's why they are not done first. In testing for HIV, the virus that causes AIDS, the enzyme-linked immunosorbent assay (ELISA) is done before Western blots and viral culture because it is cheaper; however, it is not as good.

Testing Tests

What does it mean to say that a test is "not as good"? The quality of a diagnostic test is assessed by several measures. One is *sensitivity*, or how often the test correctly identifies those with the disease. In other words, it is the probability of a positive test given that the patient in fact has the disease. Sensitivity's companion measure is *specificity*, the probability of a negative test given that the patient is in fact disease free. You can see that the calculation of sensitivity and specificity requires the use of a group of known "true positives" and "true negatives" whose status is determined from some other measure or test, presumed accurate. This other test is called the "gold standard." For example, the results of an ELISA for HIV might be assessed by the test's manufacturer as correct or not depending on whether they match the results of a definitive clinical examination for AIDS.

As a patient, however, your interest is not in the sensitivity and specificity of lab tests. You need to know about the reliability of the results from a different perspective. When your doctor says the lab test has come back "positive" or "negative," you want to know the chance that it's true (or not). Two measures are relevant here. The first is the predictive value of the positive test (the probability that you actually have the disease, given that you test positive). The second is the predictive value of the negative test (the probability that you are actually disease free, given a negative result on the test).

How can the predictive values be determined? The sensitivity and specificity are essentially fixed by the physical and chemical properties of the test, and their estimates are provided by the test's manufacturer—before marketing, it must be clear that a test replicates the gold standard fairly closely. However, the predictive values cannot be known in advance because they do not depend exclusively on sensitivity and specificity. They also depend, in part, on the prevalence of the disease being tested for—in other words, they depend to an important degree on the characteristics of the population in which the test is used.

This may seem strange, but consider the following example. We have a test for a disease, and it has a sensitivity of 0.95 and a specificity of 0.95. Thus, with either a known positive or a known negative, the test has a 95% chance of correctly declaring the status of the disease. Suppose that the true prevalence of the disease is 40% in 1,000 persons tested. Such a high prevalence may seem surprising, but remember that people tested for a certain disease usually are a subset ex-

Table 2.1. Disease status and test results in a setting with a high disease prevalence

		True Disease Status		
		Positive	Negative	Total
	Positive	380	30	410
Test Result	Negative	20	570	590
	Total	400	600	1,000

hibiting signs or symptoms leading to a clinical suspicion of the illness being present. And disease prevalence rates may indeed be that high even in some asymptomatic populations; for example, it has been estimated that the prevalence of HIV infection is greater than 50% among intravenous drug abusers in Newark, New Jersey.

The tested population can therefore be classified as shown in table 2.1. For every 1,000 persons, 400 have the disease and 600 don't. Of the 400 with the disease, 95% (or 380 individuals) are correctly called positive by the test, and 95% of the 600 without the disease (or 570 individuals) will correctly yield a negative result. However, note that 20 of the 400 persons with the disease will end up erroneously classified as negative, and 30 of the individuals without the disease will be classified as positive. These people are false negatives and false positives, respectively, but neither they nor their physicians are aware of this, because they have only the lab results to go on at present. Here's the patient's perspective: For every 410 lab slips labeled "positive" on a physician's desk, 380 will in fact be true positives, so the predictive value of the positive is 380/410, or 0.927. Thus, when the patient's lab slip comes back positive, there is a 92.7% chance that it is correct and the disease is present. As to the negative slips, 570 of them are correct, out of 590 in all, for a predictive value of the negative test of 0.966, or a 96.6% chance of being correct.

But this really does depend on prevalence, as you can see from table 2.2. Let sensitivity and specificity remain the same, at 0.95 each,

Table 2.2. Disease status and test results in a setting with a low disease prevalence

		True Disease Status		
		Positive	Negative	Total
	Positive	19	49	68
Test Result	Negative	1	931	932
	Total	20	980	1,000

and let's change the prevalence from 40 to 2%. Now 20 persons have the disease, and 980 don't. Nineteen of these 20 are correctly classified as positive by the test, and 931 of the 980 are correctly classified as negative. As you can see from table 2.2, the numbers of false positives (49) and false negatives (1) can be obtained by subtraction.

Now the predictive value of the positive is 19/68, or 0.279, and the predictive value of the negative is 931/932, or 0.999. In such a situation, a lab slip with a negative reading is almost always correct, but a positive reading is true less than 28% of the time. What happened here? The test seemed pretty good at finding positives before. The properties of the test resulting in 95% sensitivity haven't changed. The much lower prevalence of the disease means that there is relatively little opportunity for true positives to occur, even if all disease cases were perfectly categorized as positive. Furthermore, even a modest false positive rate among the much larger group of negatives will provide so many positive lab results that the resulting "pool" of positives will overwhelmingly consist of false positives.

There is a judgment implication in this: the physician (and patient) must use judgment in interpreting the results of clinical laboratory tests. Ironically, the assessment of the pre-test probability of disease, subjective though it may be, must play a role in determining how much credence to give the seemingly objective lab report. This means that the doctor should have a good idea of the type of population from which the patient is drawn (with respect to risk factors and the like). The lower the prevalence of disease in groups similar to this patient, the less likely the "laboratory positive" is to be correct and the more likely it is to be a false positive. There is also a policy implication: groups to be screened for a disease should be at high risk of the disease, in order to avoid spending time and money for a small yield of true positives (and for a large yield of false positives who will have unnecessary follow-up, expense, and worry until their true status becomes evident). Perhaps that implication seems obvious (who would screen children for cancer of the colon or prostate?). However, remember that when the United States was in the grip of the anxiety sparked by the initial awareness of the AIDS epidemic, some people suggested that everyone in the United States should be screened for AIDS. If 1% of the U.S. population were infected with AIDS, the remaining 99% of the population would provide ample numbers of false positives at even a small error rate, and the *majority* of positives detected in a national screening program might well be false ones (depending on the exact sensitivity and specificity figures). This is hardly a useful paradigm for a screening program.

If prevalence rises rather than falls, the trend is reversed: the predictive value of a positive is improved, and the predictive value of the negative becomes problematic. The numbers of true negatives eventually become small enough that even if they are all correctly classified as negative by the test, any small error rate among the overwhelming number of positives will cause a preponderance of false negatives as a fraction of all negatives. To use an extreme example, in an AIDS hospice every negative test would be a false negative test, and the predictive value of a negative would be zero.

Many measurements in medicine are not simply negative or positive readings of course, even though a decision about a patient's status is recorded in such binary fashion. Patients are said to have high blood pressure or not, to have glaucoma or not, and so forth, but in reality nature often presents a continuum of possible values for parameters while arbitrary cutoff points are used in diagnosis. At a certain point, it's important to treat high blood pressure; at a certain point intraocular pressure needs to be treated to arrest the damage caused by glaucoma. Still, knowledge of the underlying distribution of values is an important basis for medical decision making and for determining when a drug has been effective in shifting the typical values of a parameter. One of the most common and familiar underlying distributions is called the normal curve.

Celestial Distributions and the Corner Bakery

The normal curve is sometimes referred to as the Gaussian curve, or the normal curve of error, because of its origin. Carl Friedrich Gauss was one of the first to apply this distribution extensively, using it to calculate the probabilities of various outcomes. A child prodigy born in Germany in 1777, Gauss became an outstanding mathematician and astronomer. He estimated the distance, sizes, and locations of celestial objects and was using a new telescope in an attempt to produce an improved estimate of the moon's diameter. While doing so, he noticed something strange. His repeated measurements varied because of errors, yet the deviations occurred in a consistent way. Most measurements differed slightly from the average; the greater the deviation or error, the less likely it was to occur. Gauss established that the probability of errors was distributed according to the normal curve, whose shape is shown in figure 2.1. The horizontal (x) axis represents measurements and the vertical axis the relative frequency of such measurements.

The normal curve is unimodal (one humped), continuous (any x value is possible, not just certain intermittent ones), and extends in both directions without touching the x-axis. This latter property implies that the area under the curve (AUC) is never quite 0—any measurement, no matter how extreme, might possibly occur. Of course, the AUC at extreme values of x is very small. More than 99% of the AUC (more than 99.7%, actually) is included within the range of x values that runs from 3σ below the mean to 3σ above the mean.

The Greek letter σ (sigma) represents a statistical value called the *standard deviation*, or SD. It is obtained by subtracting the average (μ, mu) from each x value; this process gives a set of deviations from the average, and then we can examine these differences to see whether they might typically be large or small. After all, some types of measure-

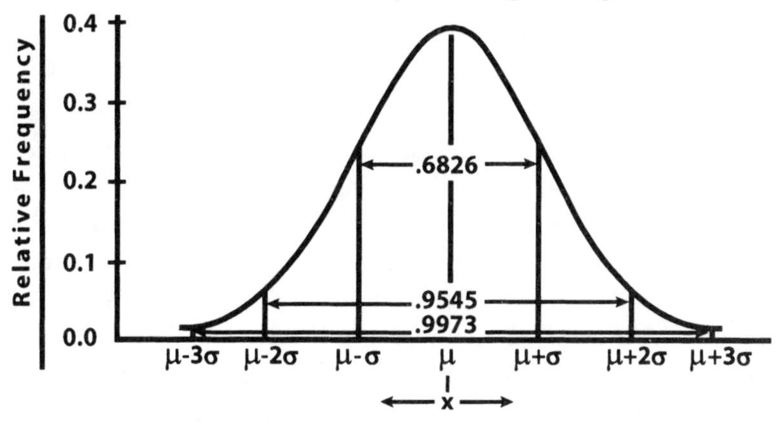

Figure 2.1. Areas under the curve of normal distribution.

ments are subject to a great deal of variability and others are relatively reproducible. It would be useful to have a number reflecting the sum of these deviations in order to quantify the amount of variability, but some deviations have positive signs and others negative, and these would cancel out each other. (The mean is right in the middle of all observations—the curve is symmetrical—so it is very easy to see that the sum would in fact always be 0.) To avoid this problem, the deviations are squared before taking their sum. (The process of taking the sum is symbolized by the Greek letter Σ, a capital sigma.) Then they are averaged by dividing by N, the size of the population of observations. Since they were squared before they were averaged, the square root is now taken to restore the scale to what it was before. The result is the standard deviation:

$$\sigma = \sqrt{\frac{\Sigma(x - \mu)^2}{N}}$$

Gauss calculated that 68% of the observations in normally distributed data lie between 1 SD above and 1 SD below the mean. About

95% of the observations lie in the range $\mu \pm 2\sigma$. And, as mentioned earlier, 3 SDs on either side of the mean encompass almost all of the AUC. More generally, Gauss discovered that if one has the mean and SD of a normal distribution, it is possible to calculate the probability of seeing an observation at any location along the x-axis.

It's not only celestial measurements that follow a normal distribution. Quality control specialists involved in checking the results of manufacturing processes often use normal distributions in analyzing the mean and variability of measurements as different as the width of aircraft parts and the amount of wine in a bottle said to contain 1 liter. There is always some variability from part to part and from bottle to bottle, so distributions are examined to see whether this variability is within acceptable tolerance limits.

Long before Gauss quantified the systematic deviations that he found, it had been understood that variability in measurements and in manufacturing processes was to be expected. Efforts were made to rein in variability by constant comparison with standard measures. Throughout history, kings had standard gold coins of various denominations and expected those produced by royal mints to match them in weight. Knowing that the match would rarely if ever be perfect, they promulgated allowable limits—excessive variation would lead to accusations of cheating. On the other hand, sometimes variability had to be taken into account in the establishment of units of measurement themselves. Many units were originally defined as distances that could be approximated using a human body. The yard, for example, started out as the distance from the tip of a grown man's nose to the tip of the longest finger on his outstretched hand. In Egyptian and biblical times, the cubit, based on the forearm, was used in building. The foot as a unit is another obvious example. Such units were useful because they were always available and required no special equipment, but obviously they were highly variable and therefore not very helpful for commercial purposes and would often lead to disagreements.

During the Enlightenment, technology and government grew to the point where a meter could be engraved on a bar of metal in Paris and serve as a standard for all of France; but even before then some people had seen the use in averaging out variability from person to person, as a way of standardizing units. In 1584, J. Koelbel suggested a way to provide a good estimate of the *rute*, or *rood* (a now-archaic unit used in measuring land), and a good estimate of the linear foot as well. He published his method in a book called *Geometry*. This was a book for surveyors rather than mathematicians: here geometry is the "science of measuring the earth," rather than a system of theorems and proofs about spatial figures as in Euclidean geometry (although such systems did originate in techniques for measuring land). Koelbel suggests taking 16 men "as they happen to come out" from church and have them line up their left feet in a single, contiguous row, one behind the next. The length of this sequence of feet would be "the right and lawful rood"; one-sixteenth of this length would be "the right and lawful foot." We see here a recognition of the importance of averaging out the person-to-person variability in some random, representative sample of foot lengths in order to arrive at some "truer" value. The true value is held to be the mean of the observations, just as in the normal distribution.

Once Gauss had demonstrated that you could graph the expected variability around observations and show how likely such deviations were, a great many uses were found for the normal curve, some trivial and some of great benefit to humanity. A great French mathematician, Jules Henri Poincaré (1854–1912), used the normal distribution to confirm his suspicion that his local bakery was cheating him. It was his habit to pick up a loaf of bread each day, a loaf sold as weighing 1 kg. Poincaré knew that one underweight loaf was not evidence of deliberate cheating if the product weighed 1 kg *on average*, because the next loaf might weigh a bit over the average. Thus, he weighed the bread he bought over the period of 1 year and found a normal distribution with a mean of 950 g. On average, he was being shortchanged by 5%, and he complained to the authorities, who gave the baker a warning. The subsequent year's data brought another com-

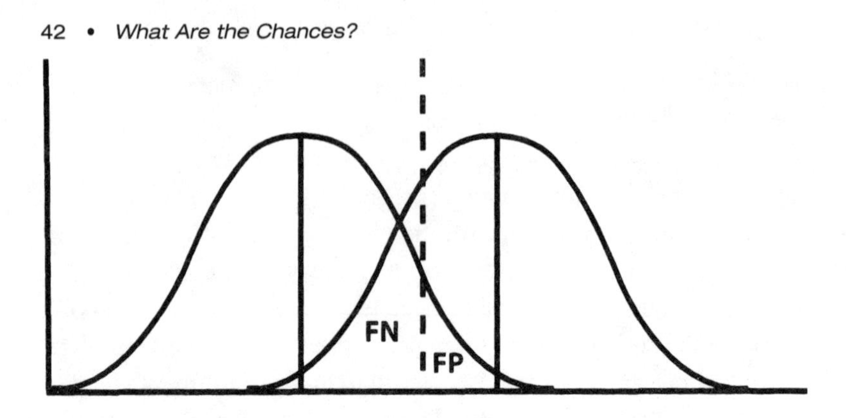

Figure 2.2. Frequency of false negatives (*FN*) and false positives (*FP*) when two normal curves overlap.

plaint from Poincaré, who said that the merchant continued to cheat. Once again the data were unimodal, with the peak frequency at 950 g. However, the distribution was no longer symmetrical. While the distribution's right-hand half was the same as before, to the left of the peak the curve was much attenuated; it was shorter in height and in length than last time. The explanation? The baker had not changed his ways except in one respect: aware that Poincaré would complain about being shortchanged, the baker always sold him the largest loaf he had on hand. When the police came calling at his shop, the baker was astonished that Poincaré had realized that the complaint was still valid.

The normal curve can help us understand why doctors using laboratory tests may still have a great deal of trouble deciding whether a given patient has a disease, even with precise, continuous measurements in hand. Figure 2.2 is a schematic representation of the problem. On the left, there is a normal curve for the distribution of clinical measurements in those without disease. It might be the distribution of intraocular pressure in patients without glaucoma, for example. The other normal curve is the distribution of measurements among people who do have the disease. These curves overlap—in nature the separation between the groups is clearly not complete.

The dashed vertical line that intersects the two normal curves in figure 2.2 can be referred to as a criterion line; if you go to the eye doctor and your intraocular pressure is above that point, you will be considered to have glaucoma. The odds certainly would be overwhelming that you'd have that disease, so it's a good test in that sense. However, there will be some disease-free people with measurements to the right of that criterion. The area that they occupy is designated FP, for false positives. There will also be some people with disease whose measurements fall to the left of that line, designated FN, for false negatives. Sensitivity is also represented in figure 2.2: it is the proportion of the area among the diseased population that is to the right of the criterion line. Specificity is the proportion of the AUC for disease-free people's measurements to the left of the criterion line.

One might suggest moving the criterion line to the left to include more of the people with disease among those who test positive. Such an adjustment would indeed increase sensitivity by correctly classifying more of those with measurements in the "diseased" curve. But it would do so at the cost of poorer specificity—a smaller proportion of the disease negative would be correctly classified as such, and false positives would increase. Move the criterion line to the right and the reverse would be true: false positives would decline and specificity would be improved. Yet, fewer people with disease would be "over the border" to the right in the positive zone as designated by the test. Thus, sensitivity would decline. In such a situation, it is impossible to pick a criterion value permitting perfect discrimination between those who do and those who do not have disease.

There is usually no single, objective, best criterion for classifying patients according to the results of medical tests. In practice, criteria for classification of outcomes depend on the consequences of an incorrect decision. Sometimes the expense and side effects of further tests and possible erroneous treatment outweigh the dangers of a mild disease. Other times it is so important to treat a disease promptly that more false positives are acceptable in lab test results, and they will be ruled out as true cases of disease later on.

"Normal" Intelligence

Many variables of primarily social rather than specifically medical importance are also normally distributed, and here, too, we find overlap between various groups of people. For example, the ability to perform various intellectual tasks is normally distributed: occasionally, we find a person who is exceptionally talented or exceptionally hopeless at something, but the extremes in level of ability are rarer than more ordinary capabilities. Those who take IQ tests end up with a numerical summary score that places them along a continuum of ability. The results are normally distributed and centered on an average of 100 points, with an SD of 15. Thus, 68% of the population has IQs between 85 and 115. The $2\frac{1}{2}$% "smartest" have IQs above $130(\mu + 2\sigma)$, while the $2\frac{1}{2}$% "dullest" have IQs below $70(\mu - 2\sigma)$.

But what does IQ, your "intelligence quotient," represent? The IQ test measures a person's skill at certain types of tasks, such as completing verbal analogies, picking the appropriate number to appear next in a sequence, or finding an optimal means of combining geometrical figures to produce another. Interestingly, when constructing an IQ test, which types of abstract reasoning questions are selected is not very important; scores on these diverse tasks are closely similar within individuals. The overall score on the pool of such questions is called *g*, for general intelligence.

In a sense, the close relationship of individual questions' scores gives us confidence that IQ really is reliable as a measure of a certain kind of innate ability in its various manifestations. Yet, in another sense, that uniformity is also a reminder that only a certain kind of reasoning is being measured (and implicitly accorded value) by the IQ test. For example, the ability to assess the emotional state of friends, to pick just the right word or action to console someone, or to sway a crowd and energize a political rally are all abilities that vary from person to person. They are surely reflections of forms of intelligence, and they are not included in the "general" intelligence index.

However, so much of schoolwork relies on the type of ability reflected in IQ scores that the scores are very good predictors of success in school, whether the outcome is measured in terms of academic marks or by number of years of education eventually obtained. IQ is a stronger predictor of such outcomes than is social class; there is a strong tendency for educational attainment of siblings, growing up together, to be different if their IQ scores are different. Conversely, genetically identical twins reared apart (as the result of adoption, for example) tend to be more similar in IQ and in educational attainment than would be expected by chance, despite different environments. However, it is not clear what predictive value IQ has with respect to more general aspects of success than these narrowly defined schooling-specific measures, because obviously there aren't any agreed-upon overall measurements reflecting success in life.

One of the most controversial aspects of IQ is the difference in the normal curves of IQ scores for African Americans and for white Americans of European ancestry. For the former group the mean value is about 85, and for the latter about 100, according to the American Psychological Association's *Encyclopedia of Psychology* (2000 edition). Given that the SDs in both groups are about 15, the mean value for whites is 1 SD above the mean value for African Americans. Thus, only 16% of African Americans have IQs above 100, whereas 50% of whites do.

We cannot scientifically conclude from the existence of this difference that it is caused by specifically genetic attributes of the two groups, however. Consider this: Among individuals raised in hypothetical identical environments, differences in IQ would clearly be genetic in origin. Now consider the converse: As psychologist Peter Gray said, "If you were raised in a typical middle-class environment and I were raised in a closet, the difference between us in IQ would certainly be due mostly to differences in our environments."

African Americans and European Americans are not genetically uniform groups, of course; they are social groups and, as such, are subject to different environments. The historic prejudice and discrimination against African Americans in the United States would be expected to leave a legacy of lower IQ scores as well as lower educational attainment. Research has shown that castelike minorities who are considered inferior by dominant ethnic groups perform on average 10 or 15 points lower than the dominant majority, in all countries where such social stratification occurs. For example, a certain genetically indistinguishable subset of Japanese people is considered socially inferior. In Japan, these *Buraku* were given civil status in 1871, when public scorning of *Buraku* was forbidden by edict of an enlightened Japanese emperor. However, even today this castelike group suffers the disadvantages of low social status, occupying predominantly menial positions; friendship with *Buraku* is greeted with contempt by the majority of Japanese. Like those of African descent in the United States, *Buraku* in Japan fare markedly less well than the majority on IQ tests. However, among Japanese immigrants to the United States, *Buraku* and other Japanese score the same on IQ tests and in school achievement—the differences disappear once pariah status and a sense of hopelessness are removed. In the United States, almost no one would know that a Japanese immigrant is a *Buraku*, and no one cares. Once the social opportunities are equalized, the normal curve of IQ for *Buraku* shifts upward and matches the normal curve for the Japanese in general.

Popcorn and the Distribution of Sample Means

The normal distribution becomes audible when you pop a bag of microwaveable popcorn. For a while, nothing happens, but then you hear an isolated pop now and then. Next you hear a few at a time. The intervals between pops get smaller and smaller until lots of kernels are popping at once with no silences between them. The number popping grows cacophonously, crests, and gradually diminishes until you again hear but a couple at a time. Finally, the action peters out and there are several seconds between pops; when the pops are few enough, you remove the bag from the microwave and enjoy your snack.

A graph of this process would look like the curve we saw in figure 2.1. The x-axis would show "time until popped," and the y-axis the number of kernels having that popping time. The left-hand tail would show that very few kernels pop right away, and then there would be a gradual rise toward the mean time. There's a symmetrical right tail too, the full length of which is usually left unexplored: when you remove the bag there are inevitably unpopped kernels.

A popcorn company might store its entire crop of 10,000 pounds of popcorn in a silo, and the crop would have a mean popping time. However, when this crop gets divided into 20,000 half-pound bags, each bag will have a somewhat different mean popping time. Each bag is a sample, and its mean is subject to sampling variability; when you pop one bag, it provides a sample mean that is an estimator of the true population mean. The mean of all the sample means has got to be the true mean, as you may realize—especially if you think about what you would hear if you were to pop all the bags at the same time. The overall mean popping time for the whole crop would be the point of maximum noise. There is a formal proof of this, known to mathematicians as the Central Limit Theorem.

We have mentioned that the probability of any observation drawn from a normal distribution can be calculated, once μ and σ are known. Similarly, sample means are also normally distributed, and their probabilities can be calculated too. Almost all sample means are quite close to the population mean, but an occasional bag of popcorn will have a bit more than its share of slow- or fast-popping kernels. The further the bag's mean from the population's mean, the more unlikely it is. Quality control personnel have an interest in making a company's product essentially uniform from one purchase to the next, so they monitor the distribution of sample means quite closely.

When a single sample is available and the population mean is not specifically known, the sample mean serves as the best estimate of the overall mean. Although it is only an estimate, it is what statisticians call an *unbiased estimate*; there is no particular reason for it to be higher or lower than the true population mean, so *on average* it will be correct. However, the element of sampling variability (which would not be present in a population average) must be present.

The larger the sample size, the less the sampling variability and the more likely the sample mean will closely reflect the population mean. After all, as the sample size approaches the population size, it includes more and more of the population; even if the sample includes some anomalous observations, these will not alter the average too much if the sample includes practically all observations. As the sample size *decreases*, however, the chance of an anomalous sample increases. Imagine a bag of popcorn with only four kernels in it. If by chance a kernel with a very short or very extended popping time gets included, it can have a drastic effect on the mean.

The effects of sample size are taken into account by creating a statistic called the *standard error*, or SE. It's the SD calculated from the data in the sample, divided by the square root of the sample size. Once this adjustment is made, all the usual Gaussian probability calculations can be performed. When the likelihood of a sample mean rather than a particular observation is to be calculated, the SE is simply substituted for the SD. For example, 68% of samples drawn from a normal distribution will have means somewhere in the range running from 1 SE below the population's mean to 1 SE above it.

Strangely enough, the samples don't even have to be from *population* data that are normally distributed in order for the *sample* means to follow a normal distribution permitting Gaussian probability calculations. The Central Limit Theorem proves this property of sample means, too, but a moment's reflection will make it seem less counterintuitive than it may at first appear. Suppose that we are looking at samples of six observations apiece, from a totally flat distribution-one forming a big rectangle when graphed. If it's a random sample, the likeliest situation is that the six are spread more or less evenly throughout the range of x. Evenly spread samples will have a mean roughly midway in the range; that is, the sample means will be approximately the same as the population mean. The rarest situation is that the random sample will consist of six observations tightly clustered at one extreme or the other, all coincidentally at very low or very high values of x. Thus, very rarely, samples may produce means that are extremely high or extremely low compared with the mean of the population from which they are drawn. Sample means that are neither spread nicely throughout the range of x nor completely bunched in the extremes will have means that differ moderately from the true population mean. These samples will occur with intermediate frequency. Hence, sample means end up being normally distributed even when the underlying population is not normally distributed at all.

Often in industrial or medical applications, the true population mean is not known, and only data on a single sample are available.

For example, testing how long lightbulbs last if they've been produced with a new type of filament may require letting a batch be illuminated until they "blow out." This type of "destructive testing" is costly, and naturally it is not desirable to test the entire output of a manufacturing plant by ruining all the lightbulbs in order to show how the true mean and the sample means compare. In other situations, involving the testing of a new medication, a sample provides the only knowledge of the cure rate prior to marketing of the medicine to the entire population. To be sure, the sample may be large and it may be representative, but it's still a sample. A useful tool called the confidence interval (CI) has been devised to permit inference about the underlying population's mean when only a sample mean is available.

Here is a statement involving a CI: "We are approximately 95% certain that the mean duration of illumination for our brand of lightbulb is between 3,500 and 3,600 hours." How is such a statement arrived at, and what does it mean? Look at figure 2.1 again, and suppose it is the distribution of sample means. Since we are dealing with samples here rather than individual observations, we can replace the SD (σ) in the diagram with the SE, and the corresponding probabilities in the diagram remain the same. Consider a specific sample whose mean is just short of 2 SEs above the population mean. For that sample—indeed, for any sample whose mean falls in that central 95%—the true population mean will be located in an interval given by the following: (the sample mean $\pm$ 2 SEs). Moreover, since 95% of samples will be in the middle of the distribution of sample means and have that property, 95% of the intervals constructed on sample means will include the population mean. Thus, there is a 95% probability that if you take a randomly selected sample mean $\pm$ 2 SEs, the interval will include the true mean of the population from which the sample was drawn. It is therefore called a 95% CI. You have 95% confidence that it contains the true mean.

Sometimes the mean is not the only measure estimated from the sample. The variability that should be represented by the SD (σ) is also unknown. In that case, the numerator of the SE is s, the *sample*'s SD, and it is used as the estimate of the population SD. In such cases, extra sampling variability is introduced into the equations—even the measure of uncertainty now has uncertainty. Another distribution, shaped similarly to the normal distribution, is used to resolve this problem. Although a bell-shaped curve, the t-distribution is somewhat "fatter" than the normal curve. This allows for the greater uncertainty: CIs need to be wider when s rather than σ is used to estimate SEs. Since uncertainty varies according to sample size, the t-distribution changes "fatness": at larger sample sizes it gets thinner. For example, when the sample size is 5, the 95% CI based on the t-distribution is given by the sample mean ± 2.447 SEs. When the sample size is 20, it's the sample mean ± 2.093 SEs, and at 30, it's the sample mean ± 2.045 SEs. In mathematical terms, we say that the t-distribution approaches the normal distribution as sample size increases. Indeed, in many experiments such as clinical trials involving large groups, the z-distribution is used even when the data provide the estimate of s for the SE; the t- and z-distributions give essentially the same CIs.

CIs do not have to be restricted to statements about 95% probability. Ninety percent and 99% CIs are also commonly used. The higher the percentage in the CI, the wider it must be. A 68% CI would be rather narrow, merely requiring a range of ± 1 SE around the mean. But who wants to have an estimate with just 68% certainty? A 100% CI, seemingly the most desirable, is actually of no practical value: it would run from negative infinity to positive infinity. To achieve *perfect* certainty about a group's measurements on any parameter, one would have to be sure to include any possible value, an infinite range. Hence, there is a trade-off when constructing CIs: greater certainty versus greater precision in the estimate.

We can never remove uncertainty from life, of course, and it can be unnerving to think that there is substantial uncertainty in such processes as the measurement of aircraft parts or the testing of drugs' effects in medical experiments. But there are patterns even in random events such as "the luck of the draw" when selecting a sample. The regularity found in these chance fluctuations can be quantitatively predicted by distributions. This is surely one of the oddest discoveries of mathematics. It is also an extremely practical insight upon which much of modern medicine, engineering, and social science depends.

The Life Table: You Can Bet on It!

The Deal of a Lifetime

On August 4, 1997, a Frenchwoman named Jeanne Calment died at the age of 122. She was a native of Arles, the town in which Vincent Van Gogh spent the year 1888. Van Gogh's productivity there impresses art historians: it was in Arles that he painted *Vase with Fourteen Sunflowers*, *Starry Night*, and more than 100 other paintings. Van Gogh made quite an impression on Jeanne Calment, too, who was 13 when they met that year in her uncle's shop. Even in adulthood she was to remember the artist as "dirty, badly dressed, and disagreeable." By the time Charles Lindbergh crossed the Atlantic in 1927, Jeanne Calment was 52 years old.

In the mid-1960s Jeanne Calment struck a deal with a lawyer that seemed mutually advantageous. He bought her apartment for a low monthly payment with the agreement that payments would cease at her death, at which point he could move in. She would thus have an ongoing source of cash to live on in her last years, and he would get an apartment cheaply, with no money down, in return for accepting the uncertainty as to when he would take possession.

After making payments for more than 30 years, the lawyer died at age 77, before she did. His family inherited the agreement: they would be in line to get the apartment, but in order to do so they would have to assume the original deal, continuing the monthly payments until she died. Her age at death exceeded the lawyer's age at death by 45 years.

Obviously it turned out that this was not a good way for the lawyer to obtain an apartment "on the cheap"; in fact, he never occupied it. However, his expectation that it was a good deal was a reasonable one, based as it was on typical human life spans. He had no way of knowing that the woman with whom he has struck the deal would have such an exceptionally long life—indeed, the longest well-documented life span on record at that time. Nor did she have any way of anticipating her own longevity, although she did feel that the abundance of olive oil in her diet—and her moderate drinking of port—could have salutary effects (an opinion that most epidemiologists would agree with today).

Individual life spans are unpredictable, but when data are collected from groups of people and analyzed en masse, very regular patterns emerge. Average life spans (called life expectancies) change quite little from one year to the next, and ages at death in a population follow a distribution with well-known and rather reproducible properties. As a consequence, the probabilities of mortality at various ages are so reliably estimated that these quantities form the basis of the life insurance industry. (In England it is called "life assurance"; the industry can assure you of the one thing certain in life and tell you when it is likely to occur.) The casino industry is also based on knowledge of probabilities, but the house "advantage" is higher for an insurance company than for any roulette wheel. And unlike the unlucky lawyer, the insurance company is not betting on one life but on millions, so the frequency of remarkable deviations can be anticipated and is not disastrous.

Insurance and Sacrilege

The lack of reliable mortality data meant that for a while the insurance industry avoided life insurance and focused on other risks. In

Rome, for example, proportions surviving at various ages had been estimated by the jurist Ulpian around A.D. 200. Although Ulpian's writings on this subject have been lost, his estimates have been preserved in the form of a passage on his method for evaluating annuities, quoted in Justinian's *Digest* (a summary of Roman law that dates from A.D. 533). These statistics were of limited utility, however, and most insurance in Greek and Roman antiquity was for goods shipped by sea, which was also the case in Babylonian and Phoenician times. Cicero had his shipment of household goods insured by private parties in 49 B.C.E. The cost of insurance depended on the underwriter's experience of losses and the value of the cargo. Some irreplaceable artworks were always considered uninsurable, and multiple losses occurring simultaneously were understood to be too large for underwriting by private persons. So that insurers would not become insolvent because of multiple shipwrecks, for example, in A.D. 58 the Emperor Claudius arranged for merchants to be indemnified against storm losses, much as a federal agency today might arrange for disaster relief beyond what insurance companies could provide.

Resistance to *life* insurance, in particular, was at a peak in the European Middle Ages. Insurance for many other purposes was available, including for pilgrims to the Holy Land: in return for a premium, the policyholder who was taken prisoner en route would have his ransom paid. Insurance based on the policyholder's death, however, was not acceptable, for Europe was now Christian. As Jacques Dupâquier wrote in his 1996 volume, *L'Invention de la Table de Mortalité*,

> Reasoning in terms of probabilities of the length of human life was inconsistent with the traditional Christian concept of death. In the closed system of medieval thought, death had a sacred character: not only could it not be the object of speculation, but it was unseemly, almost sacrilegious, to attempt to look for laws governing it. Each person's destiny was subject to the will of the Almighty, who could interrupt life at any moment, whether to repay the good by calling them near to him in Paradise or to punish the wicked with the flames of Hell and eternal damnation. This rules out any prediction, and even more so, all calculation. (my translation)

As the Middle Ages faded into the Renaissance, this attitude became less pronounced. Furthermore, social organization as well as banking, accounting, and other financial institutions grew more stable and more complex. It became progressively more feasible to offer life insurance; at the same time, the public's antipathy toward it could no longer be counted on. Thus, where a conservative ruling establishment objected to life insurance on moral grounds, such insurance had to be prohibited by law, for moral suasion was no longer enough; for instance, in 1570, life insurance became illegal in Spain, and in 1598 in the Netherlands. Yet thinking changed rapidly. Less than 100 years later, calculation of mortality probabilities and prediction of life expectancies were being established on a firm foundation of data, probability, and statistics, and no one seems to have thought it objectionable. No theologian or church official raised any questions about it or tried to prevent it. "Even better," Dupâquier writes, "the discovery of statistical regularities in human phenomena was soon to be interpreted as a new proof of the existence of a Divine Order."

Graunt's Life Table

The world was awakened to the astonishing mathematical regularities of the universe by the publication of Newton's *Principia Mathematica* in 1687. Another Englishman, John Graunt, revolutionized thinking about mathematical regularities in the probabilities of human life and death. Born in London in 1620, he started out working as a merchant, a haberdasher in particular, and was much involved in civic affairs. He held several public offices and attained the military rank of major. In 1662, Graunt published *Natural and Political Observations on the Bills of Mortality*, which was received to great acclaim. He was elected a fellow of the Royal Society in 1663, and the book's fourth edition had been printed by 1665, the year of the Great Fire of London and a year before the plague ravaged England for the last time.

The bills of mortality were weekly statements compiled from reports of parish clerks in London concerning numbers of deaths (together with distributions of ages and causes of death). Compiled in London since at least 1532 and distributed in printed form since 1625, the statements helped the government keep track of the ebb and flow

Table 3.1. Graunt's data on survivorship

Of an initial 100 living conceptions,				
at the end of	6	years	64	survive
" " " "	16	"	40	"
" " " "	26	"	25	"
" " " "	36	"	16	"
" " " "	46	"	10	"
" " " "	56	"	6	"
" " " "	66	"	3	"
" " " "	76	"	1	"
" " " "	86	"	0	"

Source: J. Graunt, *Natural and Political Observations on the Bills of Mortality*, London, 1662.

of plague and other epidemic diseases. Graunt carefully examined the quality of the data and their possible shortcomings and analyzed the data in terms of means and distributions. Then he made a particularly valuable contribution: he structured the data in a format that statisticians today call a life table. He believed that such tables would demonstrate the existence of hidden laws underlying and governing human mortality.

Rather than presenting the particulars of the numbers in the groups actually observed, Graunt showed how death would diminish an initial group of people called a *cohort.* A hypothetical "round number" of individuals was picked, to make the scale of mortality comparable among life tables, and observed mortality rates were applied to this round number. He published the data presented in table 3.1. More than half the people died before age 16.

Today we would start the table at birth rather than at conception, and different age categories would be used; but the concept of applying death rates at varying ages as you follow the experience of a conveniently sized cohort remains the same.

Gottfried Wilhelm Leibniz, Isaac Newton's fierce competitor for recognition as inventor of the calculus, made important contributions to life table analysis. Among other refinements, he figured out how to obtain life expectancy at various ages when presented with data like Graunt's. This development made possible the more elaborate life tables of today, which are very rich sources of information about various aspects of human survivorship.

Getting from Mortality Rates to Life Expectancy

What is a "life expectancy"? The meaning of this statistic, and its relation to mortality rates, should become clearer if you examine a contemporary life table. Table 3.2 is a recent life table. The data concern mortality for females in the United States in 1996 and come from the Web site of the National Center for Health Statistics, a division of the Centers for Disease Control (www.cdc.gov/nchs/data). Most countries' published life tables are a few years behind current experience. The data are based on death certificates, and it is a massive task to ensure the accuracy and completeness of a set of millions of these and then to prepare them in a computerized form suitable for analysis.

The life table is always organized with one line for each age category. Some life tables are extremely detailed and provide information by single years of age. When the age category is one year throughout the table, it is called an *unabridged* life table. Table 3.2 is abridged; the first column shows the ages for each line, from X (at which the age category begins) to $X + N$ (in which N is the width of the interval).

What is the chance of dying between ages 10 and 15? This is, technically speaking, an age-specific probability of mortality. In table 3.2 it is denoted by Q, corresponding to notation elsewhere in probability: P is the probability of surviving the interval and $Q = 1 - P$ is the complementary probability of failing to do so. These rates are calculated from death certificate data and birth dates in the population, and nothing else need be supplied in order to calculate life tables; the rest is all generated using Graunt's logic and Leibniz's refinements. It's not that simple to obtain the Q_x column, however. People who die at age X in a certain year were not all age X exactly at the start of that year. The cohort at risk of dying at X changes while the year pro-

Table 3.2. Abridged life table providing mortality data on females in the United States in 1996

X to $X+N$	${}_NQ_X$	l_X	${}_ND_X$	${}_NL_X$	T_X	E_X
00 to 01	0.00659	100,000	659	99,435	7,907,507	79.1
01 to 05	0.00135	99,341	134	397,043	7,808,072	78.6
05 to 10	0.00083	99,207	82	495,812	7,411,029	74.7
10 to 15	0.00093	99,125	92	495,426	6,915,217	69.8
15 to 20	0.00220	99,033	218	494,654	6,419,791	64.8
20 to 25	0.00242	98,815	239	493,488	5,925,137	60.0
25 to 30	0.00311	98,576	307	492,128	5,431,649	55.1
30 to 35	0.00430	98,269	423	490,336	4,939,521	50.3
35 to 40	0.00608	97,846	595	487,848	4,449,185	45.5
40 to 45	0.00858	97,251	834	484,325	3,961,337	40.7
45 to 50	0.01269	96,417	1,224	479,247	3,477,012	36.1
50 to 55	0.02036	95,193	1,938	471,421	2,997,765	31.5
55 to 60	0.03150	93,255	2,938	459,363	2,526,344	27.1
60 to 65	0.05068	90,317	4,577	440,808	2,066,981	22.9
65 to 70	0.07484	85,740	6,417	413,497	1,626,173	19.0
70 to 75	0.11607	79,323	9,207	374,780	1,212,676	15.3
75 to 80	0.17495	70,116	2,267	321,360	837,896	12.0
80 to 85	0.27721	57,849	6,036	250,275	516,536	8.9
85 and up	1.00000	41,813	41,813	266,261	266,261	6.4

gresses, as some people attain that age and others safely progress to the next one. Hence, an adjustment is made to ensure that the deaths are related to the correct numbers of person-years of risk at a given age in a particular year. The net result is that ${}_5Q_{10}$, the chance of dying between 10 and 15, is 0.00093.

Suppose that we started off with 100,000 persons. How many would still be alive at age X? This is presented in the third column of table 3.2, which corresponds to the data published by Graunt, and is labeled l_X. The 100,000 is called the *radix* of the life table; the Latin word means *root* (and is the root of *radishes*). The initial group would be diminished by 0.00659, or 659 babies, in the first year of life. The number of deaths, 659, is shown in the column labeled ${}_ND_X$. Subtracting the ${}_ND_X$ for a given interval from the l_X on the same line gives the new l_X for the next line. Here, we have 100,000 − 659 = 99,341, which is the value for l_1.

At this point, we know the number of people who start each interval, and since we know the number who die during it, we also have the number who survive it. We can now calculate an important statistic, ${}_NL_X$. This is the number of person-years lived during each interval. It is important because the life expectancy is the average number of person-years lived by cohort members, so the number of person-years in each interval is needed for the sum. To construct the ${}_NL_X$ column, the number of people starting the interval at age X is multiplied by the width of the interval. Then, since some people entering the interval die during it, the number of person-years has to be diminished a bit. A reasonable approximation would be to take off half the width of the interval for those who die during it, on the assumption that deaths are evenly distributed throughout the interval. That basic principle is followed. However, risks of mortality are not quite evenly distributed, and when distributions of *exact* ages of death are available, they are used to construct the ${}_NL_X$ column.

If you start at the bottom of the table and take a cumulative sum of the ${}_NL_X$ values as you go up, you arrive at the ${}_NT_X$ values. For example, in the last age category ${}_NL_X$ and ${}_NT_X$ are the same, but T_{80} is the sum of the ${}_NL_X$ values at age 85 (266,261) and at age 80 (250,275); thus, T_{80} is 516,536. With this cumulative sum in hand, life expectancies can also be calculated. After all, the sum T_0 on the first line is the cohort's *cumulative* number of person-years lived after age 0 (that is, years lived since birth), so we can find an average number of years that a cohort member would live. This particular average is called the *life expectancy at birth*, E_0, and is given by T_0/l_0. More generally the life expectancy at any age X is given by T_X/l_X.

Looking for Universal Laws

There is an age-related pattern in the mortality rates and life expectancies observed in table 3.2 that mirrors the pattern seen universally in life tables for large human populations. The rate in the first year of life—the infant mortality rate—is comparatively high. Indeed, the difference between this rate and those in subsequent years is so great that the first year's data are almost always presented separately even in abridged life tables. Some babies are afflicted with the hardships of prematurity or birth defects. Such problems, as well as malformations of various kinds inimical to life, become manifest and have their strongest effects in the first year. Mortality falls sharply thereafter, to a low between 5 and 15. At such ages, serious illness is rare and parental supervision renders fatal accidents uncommon. Next, youthful follies of various kinds, accidents, and driving cause a sharp uptick in ${}_{N}Q_{X}$ that is followed by regular swift increases as people age. Interestingly, in table 3.2, rates exceeding infant mortality are not seen until ages in the forties.

Specific levels and other details of life tables vary from country to country and over time. These differences are determined by social conditions such as access to proper nutrition, clean water, and medical attention. In most countries, from one year to the next there are minuscule improvements in each age-specific rate as the result of gradual improvements in the standard of living, although reversals in Russia recently have provided an important exception. Still, when looking at a particular life table, the pattern in ${}_{N}Q_{X}$ values by age, described in the previous paragraph, will be evident. The generality of this pattern throughout human societies must originate in the biological aspects of viability and aging. Graunt's expectation has been borne out by subsequent experience.

It is obvious that life expectancy should decline from interval to interval. There is, however, a strange phenomenon manifest in the E_X column in table 3.2. Look at the 60-year-old female, for example. She will live another 22.9 years on average. However, if she makes it to 65, her life expectancy is *not* 22.9 diminished by the 5 years she just got through. It is *not* 17.9 years; E_{65} is 19 years. This phenomenon is sometimes flippantly summarized by demographers in the saying, "The longer you live, the longer you'll live." It is thought to be the result of heterogeneity in frailty, coupled with selection. Since there is variability in the propensity to have a heart attack, cancer, or perhaps even an accident, those likely to suffer these events tend to experience them sooner than other people. Thus, cohort members who are less susceptible to these problems make up a disproportionate fraction of the people in successive age categories. The phenomenon is more noticeable at higher ages (probably because of the increasing importance of variability in frailty), but, unfortunately, it does not go on indefinitely.

Betting against Your Own Survival

When you buy life insurance for some fixed term such as 1 or 5 years, you are placing a bet with an insurance company. You are betting that you will die in that interval; it is betting on the vastly greater probability that you will survive. The insurance company loses its bet and pays up only if you die. The insurance company has the life tables, and it prices the policy accordingly. For example, in table 3.2 a 35-year-old female has a 0.00608 probability of dying in the next 5 years. Thus, if an insurance company charges 35-year-old females $6.08 per $1,000 worth of 5-year term insurance, it would have exactly the amount of money needed to pay the claims for this group. Of course, insurance companies sell the policies at a higher price than that. They have administrative costs such as offices and employees and a profit margin to maintain. There is also a reserve needed in the event of chance upward fluctuation in the insured sample's death rate. Nevertheless, it is fairly easy for insurance companies to estimate what they should actually charge for the insurance, because once the life table is in hand, the other components of the premium can be readily estimated. In addition, by requiring a medical examination before issuing a policy, insurance companies can insure a healthier subset rather than the entire population represented by the life table. Consequently, insurance companies seldom go bankrupt.

Chance fluctuations are naturally of great interest to the insurance industry, but they rarely threaten the profitability of life insurance. Mortality due to "acts of war" or "acts of God" is usually specifically excluded, in order to avoid massive numbers of simultaneous claims. "Acts of God" is legal language that might include, for example, a city-destroying earthquake; it does not, in this context, include deaths due to commonplace natural causes, or even epidemics. Evidently our legal system uses language in a manner far removed from the medieval conception of "acts of God."

With those exclusions in place, natural fluctuations are rarely important. Nationwide mortality probabilities are based on so many millions of lives that the occasional centenarians or clusters of deaths have but negligible effects on life expectancy. Even outbreaks of disease, including AIDS, do not affect national-scale life tables much in Western nations. (It is estimated that 0.58% of the inhabitants of North America are infected with the AIDS virus.) The trick for an insurance company is to have a group of policyholders representative of the nation as a whole, and as close to a national size as possible. The larger the insured pool and the more general it is, the less subject it will be to sampling fluctuation and to clustered deaths, and the more closely it will follow the very gradual changes in mortality observed over time.

Moreover, in the event that an improbably and unprofitably large cluster of claims comes along, insurance companies have reinsurance to protect them. Reinsurance is the insurance companies' failure insurance—they pay into a pool of money that is then available to bail out a company that would otherwise default on the payment of numerous claims. However, reinsurance is more important for property losses than for loss of life. A few unusual hurricanes can wreak billions in damage. And storm damage comes in clusters a lot more than deaths do.

If the probabilities are so thoroughly understood by the insurance companies, and so thoroughly stacked in their favor, is buying life insurance an unwise bet? No. Even though the odds of winning your bet are small, life insurance is useful when it serves the purpose of guarding your family against financial ruin. At a time when a large investment can yield a 10% annual rate of return, most financial advisers would suggest that those supporting a family have policies valued at 10 to 15 times the annual income of the insured. If the insured dies, the financial status of the family would remain essentially unchanged. Investment income on the capital provided by the insurance would roughly replace the income of the deceased. Therefore, having life insurance is a fundamental part of financial planning, and it is socially desirable to have insurance companies protect themselves effectively against bankruptcy.

The type of life insurance I have described is a straightforward kind of bet, but there are many variations. For one thing, some policyholders may want to be certain that they will have the right to renew the policy as they enter later (and riskier) age categories. As any group ages, it will increasingly contain individuals who have become ill and whom the company would like to drop, so guaranteed renewable insurance costs a bit extra.

Another variation is called decreasing term insurance. With ordinary term insurance, premiums follow the death rate and rise dramatically at higher ages, at a time in life when the expenses of housing and educating children may be past. Both changes can be accommodated simultaneously by reorganizing the policy to keep the premium constant in return for a gradually decreasing payoff.

Another product available for those insisting on a permanently fixed premium is the whole-life policy. Here the premium is constant but set at a much higher cost than warranted by the risks at the start of coverage. The excess contributes to a cash value that is returned to policyholders who discontinue their insurance; for those continuing coverage (and continuing to live) into higher age categories, the cash value is available to cover the additional risk, especially as the insurance company invests it over time. In addition, whole-life insurance offers the company additional cash in hand, useful in the unlikely event of shortfalls or for such needs as new office buildings.

Just Your Type

Some risk factors are much more widespread than, say, the AIDS virus. In such circumstances, the combination of increased risk and high prevalence may make it useful to calculate life tables separately for those in higher and lower risk categories. Insurance related to automobile accidents in particular is priced differently according to the type of individual buying the policy—age, gender, state of residence, and history of accidents all play an important role in categorizing the customer and predicting risk.

Another example concerns smoking. Many companies prefer life insurance premiums to reflect the increased costs of smokers' elevated death rates. It is also profitable to offer policies at lower cost to nonsmokers, who naturally relish the advantage of being classed with similarly low-risk people. So, you see, although I have stated that individual life spans are unpredictable, this is not *completely* true. Life tables give us averages (predictions of a sort), and they are accurate enough for betting purposes. Certainly, individuals' length of life may vary substantially from averages, but the more you know about a person's risk factors, the more you can reduce this variability, and the better your estimates of the predicted mortality rates for a specific type of individual.

The differences between risk groups can be quite substantial. Figure 3.1 shows estimated annual mortality rates by age per 100,000 males (with vertical lines representing a range of variability), based on the follow-up of about a half million men in the American Cancer Society's Cancer Prevention Study II. The bottommost line shows the lung cancer death rate for those who never smoked. Although the risk of dying from lung cancer rises slightly throughout life, it remains very small indeed for nonsmokers.

The other extreme is shown by the top line of figure 3.1, which tracks the experience of the group that smoked cigarettes continuously during the follow-up study. Their risk increases much more quickly than that among the nonsmokers. Indeed, at age 80 those continuing to smoke had a death rate from lung cancer in excess of 1,500 per 100,000—more than 1.5% per year—a rate vastly greater than that on the lowest curve. (The rates shown are subject to some sampling variability because they are based on limited numbers of deaths at a given age. Where possible, vertical bars have been inserted to depict the variability around the estimates: they are ± 1 SE in height. Interpretation of the SE as a measure of variability is discussed in chapter 2.)

Quitting smoking is advantageous compared to continuing to smoke. The intermediate curves, as you go from top to bottom in figure 3.1, represent the rates for groups of people who had quit smoking cigarettes at earlier and earlier ages. Note that the risk doesn't return to that of

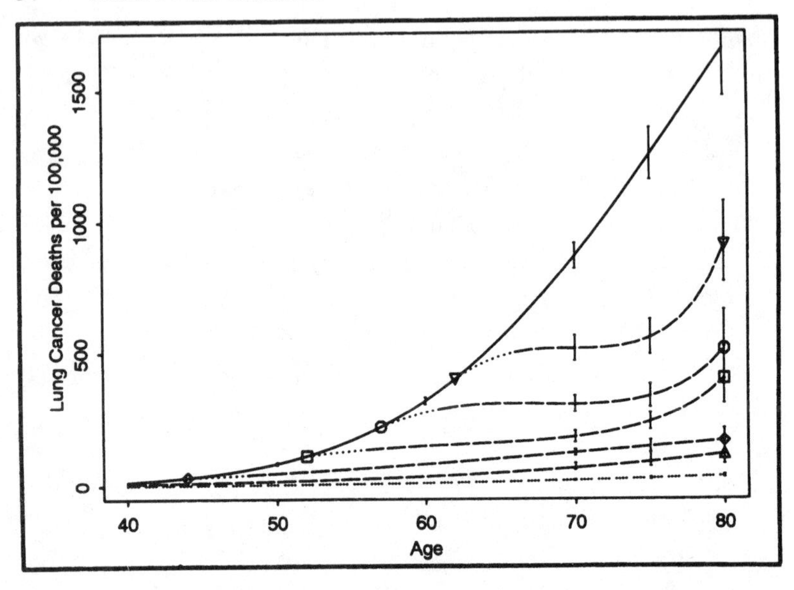

Figure 3.1. Model estimates of lung cancer death rates by age for male current (solid line), former (dashed lines), and never smokers (dotted line), based on smokers who started at age 17.5 and smoked 26 cigarettes/day. The five age-at-quitting cohorts are distinguished at the age of quitting and also at age 80 as follows: Δ, 30–39; ◇, 40–49; □, 50–54; ○, 55–59; ∇, 60–64. *Source:* Reproduced, by permission, from M. Halpern, B. Gillespie, and K. Warner, "Patterns of Absolute Risk of Lung Cancer Mortality in Former Smokers," *Journal of the National Cancer Institute* 85, no. 6 (1993): 457.

a never smoker, however. For example, the risk of dying of lung cancer at age 70 for someone who had quit smoking in his early fifties is severalfold higher than the rate for the never smoker at 70. It is also more than double the risk seen in those who had quit in their thirties. Even though the mortality rates don't decrease for those kicking the habit, smoking cessation is very important because it confers a *relative* benefit—relative to the risks sustained by those continuing to smoke.

Some lifelong smokers never get lung cancer, whereas some nonsmokers die young from it. Most people probably know someone like my Aunt Mary, who smoked heavily throughout most of her life, and died just short of her 101st birthday—but they probably don't know many such people. Despite Aunt Mary's beating the odds, many more friends and relatives who smoke heavily suffer serious health problems, or die young. Although we can't predict what will happen to individuals, we can certainly reason probabilistically about what is likely to happen. The probabilities in figure 3.1 are quite predictive *for groups*, because 1,000 heavy lifetime smokers *will* die from lung cancer at a much higher rate than 1,000 nonsmokers: they would be better off choosing to be in a low-risk group rather than a high-risk one.

Alternative Conceptions

Use of the life table method is by no means restricted to the analysis of mortality. Any event that has a distribution of occurrence over time may be analyzed using life tables, even events occurring to inanimate objects. In engineering, for example, the lifetimes of mechanical or electrical components are often the subject of study, and manufacturers offering replacement guarantees like to know in advance the typical life expectancy of their products. With this knowledge, they then can set the length of the guarantee period so that it will end before failures become numerous and repair or replacement costs climb. That's why the warranty always seems to expire just before your gadget breaks. In addition, manufacturers sell repair contracts for many appliances such as refrigerators and stereos. The prices for such repair contracts are based on the products' life tables and are set to exceed the expected failure-related payout costs for the "population" of items, in the same way that prices for life insurance policies are set. However, many people find these high-priced service contracts irksome, because the cost of several years' coverage can be a significant fraction of the cost of replacing the appliance.

Not only can life tables be used to examine patterns of events other than mortality; the events under study don't even have to be undesirable. Life tables can be used to examine waiting times to events that are much sought after, and perhaps the foremost example of such events might be conception and birth.

In infertility studies, life table analyses are commonplace. In healthy well-nourished general populations, about 85% of married couples trying to conceive will have done so after 1 year. In the second year, roughly an additional 10% will conceive. After that, there is usually some medical reason why pregnancy has not yet occurred (although sometimes it's just a long run of bad luck). The most fertile couples are the most likely to conceive at once. In the case of mortality, it was said earlier, "The longer you live, the longer you'll live," but for fertility it's a case of "The longer you wait, the longer you'll wait." As time goes on, the group that has not yet conceived includes an ever-increasing proportion of completely infertile people (and those with extremely long waiting times for conception).

Not only are conception probabilities studied with life tables, but the effects of treatments intended to increase such probabilities are analyzed with the method as well. For example, one possible reason for infertility is an inadequate production of progesterone. In some cases, fertilization is successful, yet progesterone deficiency makes it impossible to retain the fetus. The deficiency can be overcome by using progesterone suppositories. How do doctors know this? The evidence in favor of this therapy came in the form of life tables such as table 3.3. It demonstrates that progesterone supplementation can result in fertility levels that approach or exceed normal levels. Treatments for infertility are routinely assessed using life tables.

Table 3.3 also demonstrates the effects of small, specially selected samples. The cohort in this study consisted exclusively of people known to have infertility resulting from a low progesterone level, and the number of patients studied was small, fewer than 100. A cumulative pregnancy rate of 100% after less than 1 year is not realistic for fertility clinic patients in general: the sample size here precludes accurate estimation of the small percentage for whom this treatment wouldn't work (and not all infertility clinic patients have this particular problem).

Demographers also use life tables to analyze the impact of *social* factors on conception probabilities. For instance, social scientists have

Table 3.3. Life table analysis of data from patients treated with progesterone suppositories

Time (months)	Cumulative Percentage Pregnant by End of Month
0	5.7
1	20.0
2	28.6
3	31.5
4	37.2
5	48.6
6	60.3
7	69.9
8	83.3
9	91.7
10	100.0
11	100.0

Source: D. L. Murray, L. Reich, and E. Y. Adashi, *Fertility and Sterility* 51, no. 1 (1989): 35. Data reproduced by permission of the American Society for Reproductive Medicine.

noticed that in poorer, more rural societies, use of contraception is low and large families are desired, often because of a mixture of traditional religious cultural values and the need for farm labor. At the same time, infant mortality is high. This led to the expectation that under such circumstances a woman will try to get pregnant again more rapidly if a child dies young. Is this "child replacement hypothesis" supported by data?

The data presented in table 3.4 are derived from the Nepal Fertility, Family Planning, and Health Survey of 1991. The first column shows time elapsed, in intervals, since giving birth. The other two columns show the corresponding probabilities of giving birth to a subsequent child. The probabilities are calculated separately for those whose previous child died and for those whose previous child remained alive. At every elapsed time interval, the probability of giv-

Table 3.4. Cumulative probability of subsequent birth

Months since Birth of Index Child	Index Child	
	Alive	Dead
0	0.0000	0.0000
9	0.0035	0.0228
12	0.0174	0.1135
15	0.0380	0.2006
18	0.0667	0.2679
21	0.1159	0.3466
24	0.1888	0.4785
27	0.2571	0.5571
30	0.3253	0.6145
33	0.3914	0.6693
36	0.4636	0.7237
39	0.5239	0.7548
42	0.5747	0.7795
45	0.6199	0.7998
48	0.6549	0.8299
51	0.6891	0.8518
54	0.7164	0.8617
57	0.7538	0.8755

Source: Sameer Rajbhandary, "Fertility and Child Mortality in Nepal: An Econometric Analysis," University of Colorado Department of Economics, Ph.D. dissertation, 2000, p. 61. Data reproduced by permission.

ing birth again is higher if the previous child died rather than survived. Thus, the child replacement hypothesis is supported by these data, as it generally is in studies of this type.

* * *

When we look at statistics concerning birth and death rates, we might be struck not only by their regular, predictable patterns but also by the relative frequency of the underlying events they describe. A young woman in the childbearing years is much more likely to give birth in a particular year than die in it. Even death, however, is not that rare an event. In table 3.2, the lowest death rate seen is that for children 5–10 years of age: it is 83 per 100,000, or the better part of 1 in 1,000. A person who buys a lottery ticket hoping for a jackpot has a vastly greater chance of dying in the year of purchase than he does of claiming that top prize, an unsettling observation for anyone who gambles. However, as we come to examine rarer and rarer events, such as lottery jackpots and other occurrences with extremely small probabilities, we find a great increase in sampling variability and therefore increased difficulty in prediction, especially as compared with the regularity of the life table. Moreover, the interpretation of clusters of events that are supposedly extremely rare becomes especially problematic. Special statistical techniques have therefore been developed for the rarest events, which I discuss in the next chapter.

4

The Rarest Events

Isn't It Amazing?

I was once spending a pleasant evening at home when a terrible thought suddenly crossed my mind: my grandmother had just died. It's hard to explain why, but I was struck by this perception and instantly convinced of its truth. Sure enough, a moment later, the phone rang. With heavy heart, I picked it up and steeled myself to receive the bad news; instead, it was a repairman. The phone call had nothing to do with my grandmother, who lived on for many years.

Such a story rarely gets repeated (thankfully). It is not remarkable. Stories are remarkable if they involve startling sequences like the tale of someone who imagines or dreams he or she is about to receive notification of a death or a prize, and does so immediately thereafter. Those sequences are not only considered worthy of recounting, they are sometimes held to be evidence of psychic powers, or of supernatural beings providing information to favored individuals.

Such explanations are not necessary. After all, on a given day many thousands of people are struck by the thought that a certain event may be about to happen. Those events that actually *do* occur make up a very tiny proportion of all such thoughts. Some of these thoughts must turn out to be accurate, due to chance alone. Fear of a relative's death is widespread, so on any given day, death is bound to take some small fraction of the relatives of those who fear its occurrence.

The same principle holds true for dreams. On any given night in a large city, people have millions of dreams. Lots of dreams are nonsense, but people do dream about known individuals and possible events. It should hardly be surprising that some matches with reality do occur, given a staggering number of both dreams and potentially matching events from which to choose (any one of your acquaintances might buy a new car, get engaged, change jobs, or win at the racetrack). However, it is difficult to assess the *rate* at which matches with reality occur, because the reports that reach your ears tell only of dreams that came true, while matches are only a part of the story. Nonmatches are vastly more frequent, along the lines of "I dreamt I met my old friend, but I did not." But the boring stories that reflect no coincidence do not get retold, while the person who has a dream that *does* come true recounts this experience. Such dreams tend to send a shiver down the spine because they are often regarded as a demonstration of "psychic powers." It is a much less common reaction to comment, "Out of thousands of dreams I have had in my lifetime some rather predictable ones came true, as we might expect." People do not like to think of their own "amazing experiences" as unremarkable coincidences.

Sometimes sequences cannot be dismissed as coincidence, since cause-and-effect associations are present no matter how unlikely it may seem. A case in point is the phenomenon of voodoo death. Western medical workers who do not subscribe to the voodoo belief system have observed the casting of a spell and the subsequent sickening and death of the "hexed" individual and have documented these observations in the medical literature. The mortality probability for any given individual is ordinarily very small, but it is high among the hexed, so something other than random chance or coincidence must be at work. For example, an article published by Kenneth Golden in the *American Journal of Psychiatry* in 1977 presents this case:

> A 33-year-old Black man from a rural area near Little Rock, Ark., was admitted to the neurology service of the University of Arkansas for Medical Sciences. The patient had been having seizures recently and had become increasingly irritable and withdrawn from his family. When he could no longer be detained safely on the neurology service, he was transferred to the psychiatric ward, where he became increasingly more agitated, confused, and almost delirious. He became very fearful when people approached him, and he began to hallucinate. He finally slowed down after being given 1000 mg of chlorpromazine (Thorazine), but the necessity for bed restraint remained. All neurological findings, including a brain scan, proved normal.
>
> After two weeks of hospitalization the patient suffered a cardiac arrest. All efforts to revive him failed. An autopsy provided no reason for the death.

The patient's wife, however, did provide a reason. He had angered a woman whom their community believed to be endowed with the power to cast fatal spells, so, according to their beliefs, his fate became inevitable.

In 1992, another article published in the United States by Dr. C. K. Meador in the *Southern Medical Journal* described what two medical doctors had observed. A man who believed himself to have been marked for death by a voodoo priest during an argument had stopped eating and had become so weakened that he required hospitalization. He was being fed through a tube and was in a stupor near death. Doctors found no organic disease.

In front of the patient's terrified wife and relatives, the doctor in charge of the case "revealed" to the patient that he himself had recently had a violent argument with the voodoo priest about the patient, and that under dire physical threats from the doctor the voodoo practitioner had divulged the nature of the patient's problem. Thanks to voodoo, a lizard inside the patient was consuming all his food and his guts as well. After this explanation, the doctor then administered an emetic injection, and through sleight of hand produced a lizard at the end of the vomiting. The patient fell asleep, woke the next morning with a ravenous appetite, and was discharged in a week.

These illnesses, and other similar conditions in the medical literature, have a physical cause. They are the result of the physical effects of tremendous mental stress. Various specific mechanisms have been suggested by neurophysiologists, especially for the fairly numerous reports of sudden cardiac death following voodoo spells. In any case, it is obvious that extreme fear can cause a fatal heart attack, as we can see from newspaper accounts of such deaths among elderly robbery victims and hostages. Clearly not coincidental, the deaths following voodoo spells are caused by some form of exaggerated, negative placebo effect rather than by the direct mechanisms claimed by voodoo practitioners. The evidence for the power of suggestion as the origin of physical problems among the "hexed" is strong: voodoo death only occurs if the victim and all his or her friends and relatives believe in the power of the voodoo priest to cause it. They must all believe that the particular spell will be fatal and treat the victim as they would treat any individual facing imminent death, in order for the illusion to be complete and effective. Nonetheless, there's a physical explanation here. The reasoning is not limited to *post hoc ergo propter hoc*—"after it, therefore because of it."

The Statistics of Surprises

There are many situations in which the mechanism causing a certain phenomenon is entirely unknown or is strongly disputed. Then it becomes difficult to distinguish between chance and causality as the source of the confluence of events based solely on extreme rarity. A classic image of a surprising rare event supposedly arising solely from the operations of chance involves monkeys hitting the keys of a keyboard. My computer keyboard has 55 keys. A monkey hitting at random would have approximately a 1/55 chance of hitting the key needed to type a given character. (I say approximately because the keys are not all the same size, some locations may be easier to hit, and so on.) My word-processing program informs me that the previous chapter of this book had 41,400 characters. Thus, the chance is $(1/55)^{41,400}$ that a given series of 41,400 random keystrokes would type out the chapter. That's a very small probability, but that probability isn't 0 either; had we but monkeys enough and time, we would expect that eventually the sequence of characters forming that chapter would be struck at random (a humbling thought for any author).

An implication of this example is that you could come across a text and judge it to be the result of human agency, when in fact it is the result of a random sequence of events. However, this implication is obviously not very realistic. When you read a published work you know that it wasn't generated by the random hitting of letters or a random selection of words that then coincidentally formed a interpretable sequence. (One could quibble. The Dada artistic and literary movement, which flourished in Paris around 1920, did produce some examples of poetry composed by drawing words at random from a bag. In addition, some readers find the later works of James Joyce very close to randomly generated, although there is some question as to whether they are in fact interpretable.)

Realistically speaking, you know that a readable book is not a product of randomness. You know that because you also know a lot about monkeys and people, not just about probability. The real problem arises when the example is not contrived, something highly improbable is observed, and your knowledge is essentially limited to the probabilities in the face of great uncertainty as to mechanisms of causation.

For example, suppose there is a cancer that does not run in families. It occurs apparently at random ("sporadically," as doctors say) and strikes $\frac{1}{2}$ of 1% of the population—it's a rare disease. Yet, in a single family living next to a waste dump site, three children get the cancer. This seems unlikely by chance and might lead us to say that living next to the dump was risky. After all, in a set of three siblings selected at random, the probability that all three would get the cancer is 0.05^3, or 1 in 8,000, which surely meets most people's definition of an event that is unlikely by chance. However, in a country the size of the United States, more than a million families have three children. We'd therefore expect 1 in 8,000 of these families to have all three children get the cancer—that's 125 families out of every 1 million.

These little clusters of three cases apiece are expected on the basis of chance alone. If these 125 families are distributed in the same way as any other 125 randomly selected American families, many will have homes in nice locations, but a few will live near dumps, high-tension electrical wires, chemical factories, and the like. Those that do will consider only their own experience as informative. The family home is near a dump site and the children have a rare cancer, so what could be more "obvious" than a connection between the exposure and the disease?

Epidemiologists know, however, that when small probabilities are applied to vast denominators, astonishing coincidences may arise even in the absence of any cause-and-effect association. Judging whether the association is in fact causal requires knowledge not just of selected instances, but of the entire pattern of disease throughout the population. Thus, if the rate of "three siblings with disease" is 1/8,000 families in general, but 1/800 for families living near dump sites, we'd say that the relative risk is 10-fold higher in those with the exposure. That would not be *proof* of a causal association, especially in the absence of any known mechanism, but it would certainly be good evidence compared to a single selected instance. This kind of comparative evidence is required because in the total *absence* of any association, the rates would be 1/8,000 in *both* the exposed and unexposed segments of the population, yet a family with three cancers would still be impressed by the "association" between the disease and whatever looked suspicious in their vicinity. Their reported "association" would be called *anecdotal evidence*, a term sometimes considered demeaning by those who present such observations. Nonetheless, it is a form of evidence whose greatest value may be to raise questions that can be answered by scientific comparisons involving numbers or rates of cases that arise in various groups. By itself, an anecdotal report provides inadequate information about the risk conferred by an exposure; at worst it is misleading, and at best it is inconclusive evidence.

Among those who are not scientifically trained, there is much resistance to this type of conclusion (or nonconclusion). First, some people harbor a faith that scientists should be able to make some judgment based on whatever evidence is available, no matter how sparse the information; "cannot be determined" is often not considered an acceptable answer. Second, people don't like to have their experience held in little esteem. But there is another key reason why some people abhor the skeptical attitude toward anecdotal reports of "surprising" clusters of events. They are unfamiliar with an important phenomenon that arises from the workings of probability: randomness comes in bunches.

If it seems counterintuitive that randomness comes in bunches, imagine looking up at the night sky and seeing the stars arranged in an absolutely regular grid. The stars form the corners of the squares in this grid, so the sky has the appearance of thousands of boxes at regular intervals. You would be astonished to see this, and it would be difficult to imagine a less random arrangement. A random distribution must be the opposite of this grid in a certain respect: it would have some empty areas and some clumps, *not* a clump-free pattern. Only a perfectly regular grid is completely clump free.

It's pretty obvious that assigning random locations to disease cases would create some areas where no disease occurs: a patchy, scattered appearance would be evident on a map of random locations. And the complementary fact is that places on the map where no disease occurs must be balanced by some places where many cases occur. It is poor science to pluck out these local concentrations and throw out the information on the empty regions—and then regard the isolated concentrations as amazing and rare. The degree to which concentrations of disease are surprising or not must be a function of the context giving rise to the entire pattern of disease.

Why must the entire context be taken into account, and not just the rarity of getting many cases together within (for example) a family of a given size? Consider this analogy: the probability of getting 10 heads in a row when tossing a coin is less than 1 in 1,000, so it would be a surprising result—but it's only surprising for an individual sequence of 10 tosses. If you pick up a coin and toss it 10 times, it would indeed be amazing to get all heads. However, if you have millions of 10-toss trials, you know in advance you'll have thousands of trials where the result is 10 heads. In such circumstances, it is simply incorrect to pick out a trial with 10 heads and insist that there must be a reason for the 10 heads other than mere chance. It's just the effect of enormous denominators, in this case millions of trials of size 10.

There is a further problem in deciding whether a cluster of disease has a specific underlying biological cause, and has not occurred by chance. Suppose that the disease rates in two groups being compared actually *are* different, but not by much. The rates might be 1/8,000 in one group and 1/8,001 in the other. Are these different? How about 1/8,000 versus 1/8,100? Where do we draw the line and conclude that we are observing different rates?

The line should be drawn when the difference exceeds the fluctuations we would expect due to chance alone. As you know, there are chance fluctuations in rates. Even if tossing a coin had a true outcome probability of 50-50 on average, a particular set of 10 tosses might well differ from this, though usually not by much. The effects of chance fluctuation on rates are exacerbated when the rates are very small. A single additional case of disease may raise a rate from 1/10,000 to 2/10,000, a small increase in absolute terms but a doubling of the relative risk. It won't do to say we'll ignore the relative risk and only consider a whopping big increase in absolute risk as meaningful, since few exposures in real-life epidemiological studies engender very large disease rates in absolute terms. Even cigarette smoking, which confers roughly a 10-fold relative risk of lung cancer compared with the risk among nonsmokers, elevates the absolute rate of lung cancer mortality to roughly 1/100.

The crash of the *Concorde* supersonic aircraft provides a catastrophic example of the extreme variability inherent in rates of rare events. Before the crash on July 25, 2000, there had never been a fatal accident involving the *Concorde*, so the observed rate of such accidents was zero, and by that measure the plane was the safest in the skies. The Boeing 737 series had 0.33 fatal accidents per million flights, with 31 million flights to its credit. The Airbus 320 had 0.55 fatal accidents per million flights, with 7.3 million flights. Note that aircraft safety engineers measure these failure rates on a *per-flight* basis rather than on the basis of passenger miles traveled, because by far the riskiest parts of an airplane trip are the takeoff and landing, and each flight has a single pair of these no matter how long the flight may be. In addition, from an engineer's point of view, the rates of reliability or mechanical failure for a plane should be independent of the number of passengers carried, although from a liability point of view that number matters a great deal.

After the crash, the *Concorde* became the plane with the worst safety record. There are not many of the supersonic aircraft (about a dozen in regular use), and they make relatively few flights. By the year 2000, there had been roughly 80,000 *Concorde* flights. One crash elevated the rate from 0 to 1/80,000, or 12.5 per million flights. When extremely rare events are studied, especially in fairly modestly sized denominators, rates may seesaw because they are subject to tremendous sampling variability. In short, small rates are estimated imprecisely.

The Poisson Distribution

One distribution in particular is quite useful in dealing with this problem of chance fluctuations when rates are very low. We can use it in large populations even when the denominator is not known precisely. All we need to know is the number of events that occur (such as cases of disease or numbers of meteors striking Earth). Then the mathematics of the Poisson distribution can be used to determine whether this observed number is statistically different from some other typical expected number, above and beyond chance fluctuation. This distribution is named after Siméon Denis Poisson, a French mathematician who lived from 1781 to 1840 and was a student of Pierre-Simon Laplace (about whom more will be said later).

The Poisson distribution is closely related to the binomial distribution, which was discussed in chapter 1. Use of the binomial requires the probability of an event or hit occurring in any given trial, and the number of trials; from this information the probability of various outcomes in the set of trials can be calculated. Our interest at present is in very rare events, again produced by a probability (p) applied to numbers (n) of trials. The expected value or mean, m, in a Poisson distribution is just the same as it is for a binomial: $m = n \times p$. In fact, if you hold m constant, while reducing p to a tiny number and greatly increasing n, the binomial distribution becomes the Poisson distribution. In terms of the calculus, the Poisson is a special limiting case of the binomial as $n \rightarrow \infty$ and $p \rightarrow 0$.

Table 4.1. Probability of observing no "hits," when the Poisson expectation is that one "hit" will take place

n	p	$P(1, 0)$
5	0.2000	0.328
10	0.1000	0.349
50	0.0200	0.364
100	0.0100	0.366
500	0.0020	0.3675
1,000	0.0010	0.3677
5,000	0.0002	0.3678
10,000	0.0001	0.3679

A special problem in dealing with rare events is that even in fairly large samples, the event may sometimes not be observed at all. It is then especially difficult to estimate the underlying probabilities from a set of observations. It is here that the Poisson distribution becomes particularly useful, for calculations may be made based on m in the absence of specific knowledge of p and n. Table 4.1 illustrates why this is so and also demonstrates that the Poisson distribution is the limit of the binomial.

Suppose that $m = 1$. Various combinations of p and n would result in this expected value. A few such combinations are shown in table 4.1, together with the probability of seeing no cases or "hits" in increasingly large denominators. How often will the rare event be absent from the sample?

The heading of the last column is read, "the probability of observing 0 events in a distribution with $m = 1$." This number is simply $(1 - p)^n$, since we are estimating the chance of *nonoccurrence* in all n successive trials. Note that changes in $P(1, 0)$ occur more rapidly at first, then slow down at large n as a limit is approached. The limit is called $P_p(1, 0)$: the Poisson probability. $P_p(1, 0)$ is equal to e^{-1}.

Why is e^{-1} the limit of $P(1, 0)$ (in which e is the base of natural logarithms)? In table 4.1, $np = 1$ so $p = 1/n$. Thus, we could rewrite the probability of zero successes, which we called $(1 - p)^n$, as $(1 - 1/n)^n$. But e, by definition, is the limit of $(1 + 1/x)^x$ as $x \to \infty$. Substituting $-n$ for x and changing signs accordingly, it is an equivalent statement to say e^{-1} is the limit of $(1 - 1/n)^n$ and thus of $(1 - p)^n$.

This observation can be made more general. When m is not 1 but some other expected value, the limit is for $P(m, 0)$ is e^{-m}. After all, values of m can be altered by keeping the arbitrary sequence of p as shown, and changing the n. Suppose that m is now 2, for example, implying that all the n values are doubled. This time we substitute $-2n$ (rather than $-n$) for x in the formula defining e. Making the same changes in sign as before, we obtain a limit of e^{-2}, and more generally, the limit is e^{-m}.

In chapter 1, we saw the binomial formula

$$P(r) = \frac{n!}{r!(n-r)!} p^r q^{(n-r)}$$

in which r is the number of hits or successes for which the probability is being estimated. We can adjust this formula to obtain Poisson probabilities very easily, including those for other outcomes than zero hits. We can take advantage of the fact that we are using tiny probabilities operating on huge samples. That is, in an incredibly large number of trials (n), the number of successes (r) is extremely small. Just as $1{,}000{,}000 - 1$ is still roughly a million, $n - r$ is still roughly n. In such circumstances, the binomial term $p^r q^{(n-r)}$ is essentially $p^r q^n$, especially when q, the chance of failure, is so close to 1. But we have also just shown that $q^n = (1 - p)^n = e^{-m}$. Hence, that term $p^r q^{(n-r)} = p^r e^{-m}$.

The fractional part of the formula also gets simplified due to the relative sizes of n and r, cancellations, and the equivalence of $m = np$. The entire binomial formula at the Poisson limit becomes $P_p(m, r) = m^r e^{-m} / r!$. This is extremely easy to use, as we shall see, and it is preferable to the full binomial formula. For one thing, the binomial formula would need factorials for the population-sized n, and n can be in the millions. Exact values for such factorials can be hard to obtain.

However, obtaining values for huge factorials is not impossible. There is a useful method for approximating $n!$ when n is large, given by Stirling's formula. The method was actually first suggested by Abraham de Moivre, who published it in 1730 in *Miscellanea Analytica*, a book on probability. James Stirling also described the method in his own 1730 book, *Methodus Differentialis*, using an equivalent but more convenient formula. The latter equation is mentioned in the 1738 edition of de Moivre's book and is properly attributed to his contemporary British colleague. With Stirling's formula (and almost any calculator) in hand, $n!$ is obtained as $[(2\pi)/(n+1)]^{1/2}e^{-(n+1)}(n+1)^{(n+1)}$. The approximation is quite good and improves with increasing n: as n approaches infinity, the proportion of error approaches $1/(12n)$.

The odd ubiquity of e in mathematics is worthy of mention here, and many have felt it worthy of wonder. We have seen its important role in the relationship between the binomial and Poisson distributions, and in Stirling's formula for factorials. Perhaps the most familiar application of e involves the growth of money at a compound interest rate, r, over a specified time period, t: the amount at the end of the period is equal to e^{rt}. Radioactive decay occurs in a pattern inverse to the growth of money at compound interest, namely e^{-rt}. Eli Maor's *e: The Story of a Number* explains the applications and history of this quantity that plays a role in so much of mathematics.

In any event, the Poisson limit function is the easiest way to calculate the probabilities of rare events. Here is an important example of its application, which looks at the effects of vaccination. Suppose that 100 persons were inoculated against a disease that ordinarily afflicts 1 person per year in a population that size (say, a group of dormitory residents). Thus, $m = 1$. In the year after vaccination, none of the 100 persons contracts the disease (thus, $r = 0$). The Poisson probability of no one getting the disease, by pure luck in the absence of the vaccine, is $P(1, 0)$ which is

$$P_p(1, 0) = m^r e^{-m}/r! = 1^0 e^{-1}/0! = e^{-1} = 0.37$$

In other words, if you expect to find on average 1 case in 100 persons, 37% of samples of 100 will contain no cases, simply because of the sampling fluctuation when dealing with this small rate. Therefore, it's not really surprising to find no cases in a particular year. Now suppose that the experiment is repeated in a larger sample, a population of 1,000. With roughly the same expected proportion of cases, the value for m would be 10. Suppose that in this group, too, no cases are observed. Then we have

$$P_p(10, 0) = m^r e^{-m}/r! = 10^0 e^{-10}/0! = 0.000045$$

Thus, the finding of no cases in 1,000 vaccinees is very surprising if the vaccine has no effect. It would be fairly likely that "1/100" would fluctuate downward to no cases at all, as the result of a single person not getting the disease; however, when it comes to a population of 1,000, the chances are 45 in 1,000,000 that all 10 expected cases would just not get the disease. Indeed, sample sizes for clinical trials of vaccines are set in advance using this kind of logic, so that the effect of the vaccine (if there is one) will exceed the expected sampling fluctuation and hence be detectable.

The great thing about the formula is that exact knowledge of n and p are not required; only m need be assumed as long as the population is relatively constant. The practical advantage is that in large populations that do not change quickly over time, expected fluctuations in r from year to year can be predicted and distinguished from "real" changes in frequency—those with a specific cause. The power of the Poisson distribution goes well beyond medical applications. For example, armies often are fairly constant in size from year to year, unless, of course, a war breaks out. Unfortunately, some soldiers die even in peacetime. In the nineteenth century, soldiers in the Prussian cavalry would occasionally be accidentally kicked to death by horses. In the late 1800s, the cavalry was divided into 16 different corps. Across all years and corps, the value of m was 0.7: less than one death occurred on average. Rarely, more deaths did occur. Three deaths were observed in 11 corps in some particular year; even 4 deaths were observed (on two occasions). These were not considered excessive numbers of deaths in the sense that the deaths were the result of poorly disciplined troops being careless around horses. Numbers of deaths occurred with essentially the same frequency as predicted by the Poisson

	75	76	77	78	79	80	81	82	83	84	85	86	87	88	89	90	91	92	93	94
G	—	2	2	1	—	—	1	1	—	3	—	2	1	—	—	1	—	1	—	1
I	—	—	—	2	—	3	—	2	—	—	—	1	1	1	—	2	—	3	1	—
II	—	—	—	2	—	2	—	—	1	1	—	—	2	1	1	—	—	2	—	—
III	—	—	—	1	1	1	2	—	2	—	—	—	1	—	1	2	1	—	—	—
IV	—	1	—	1	1	1	1	—	—	—	—	1	—	—	—	—	1	1	—	—
V	—	—	—	—	2	1	—	—	1	—	—	1	—	1	1	1	1	1	1	—
VI	—	—	1	—	2	—	—	1	2	—	1	1	3	1	1	1	—	3	—	—
VII	1	—	1	—	—	—	1	—	1	1	—	—	2	—	—	2	1	—	2	—
VIII	1	—	—	—	1	—	—	1	—	—	—	—	1	—	—	—	1	1	—	1
IX	—	—	—	—	—	2	1	1	1	—	2	1	1	—	1	2	—	1	—	—
X	—	—	1	1		1	—	2	—	2	—	—	—	—	2	1	3	—	1	1
XI	—	—	—	—	2	4	—	1	3	—	1	1	1	1	2	1	3	1	3	1
XIV	1	1	2	1	1	3	—	4	—	1	—	3	2	1	—	2	1	1	—	—
XV	—	1	—	—	—	—	—	1	—	1	1	—	—	—	2	2	—	—	—	—

Jahres-ergebnis	Zahl der Fälle, in denen das nebenstehende Jahresergebnis	
	eingetreten ist	zu erwarten war
0	144	143,1
1	91	92,1
2	32	33,3
3	11	8,9
4	2	2,0
5 u. mehr	—	0,6

Figure 4.1. Original data cross-classifying mortality data by years (1875–1894) and regiments. The upper panel shows the numbers of soldiers accidentally kicked to death by horses. The lower panel shows the number of regiments that had particular numbers of deaths, and number of such deaths (right-hand column) that would be expected according to Poisson distribution. *Source:* L. von Bortkewitsch, *Das Gesetz der Kleinen Zahlen*, Teubner, Leipzig, 1898, p. 24.

distribution, so they were consistent with the sampling fluctuation expected in small rates (see figure 4.1).

During World War II, the London Blitz provided an example of "An Application of the Poisson Distribution." R. D. Clarke's 1946 article by that title in the *Journal of the Institute of Actuaries* described the pattern of damage by German V-2 bombs. He noted: "During the flying-bomb attack on London, frequent assertions were made that the points of impact of the bombs tended to be grouped in clusters," rather than to show a random distribution over the metropolis. It was as if certain areas were specifically targeted, or as if bombs

Table 4.2. Numbers of V-2 bombs on squares in a London grid, compared with the Poisson expectations.

Number of Flying Bombs per Square	Expected Number of Squares (Poisson)	Actual Number of Squares
0	226.74	229
1	211.39	211
2	98.54	93
3	30.62	35
4	7.14	7
5 and over	1.57	1
Total	576.00	576

Source: R. D. Clarke, "An Application of the Poisson Distribution," *Journal of the Institute of Actuaries* 72, no. 355 (1946): 481.

tended to fall together for some mechanical reason. The former explanation seemed to imply an unexpected precision in the control of the flight paths, and the latter explanation seemed strange, given that each bomb traveled across Europe atop its own rocket. To determine whether clustering really took place, "144 square kilometers of south London [were] . . . divided into 576 squares of $\frac{1}{4}$ square kilometer each, and a count was made of the number of squares containing 0, 1, 2, 3, . . . , etc. flying bombs. Over the period considered the total number of bombs within the area involved was 537. The expected numbers of squares corresponding to the actual numbers yielded by the count were then calculated from the Poisson formula." Table 4.2 presents the actual results.

The most common outcome was to receive no bombs at all, followed by a single bomb, but certain squares received very large numbers of hits in a real and unfortunate sense. Yet, despite the impression of clustering, clusters were no more common than one would expect by chance. The observed numbers of beleaguered squares receiving four hits, or more than five, were almost identical to the numbers predicted by the Poisson distribution, and there was scant difference elsewhere.

Making an Unlikely Event a Certainty

While people naturally try to avoid certain rare events such as contracting peculiar diseases, being kicked by horses, and being hit by airborne bombs, there are other rarities that are eagerly sought after, ones that aren't necessarily covered by Poisson probabilities. Few rare events excite as much fervent hope in so wide a segment of the population as the choosing of a winning lottery number. The odds are minuscule of winning a large prize, of course, but millions of people play, perhaps for fun and amusement, or perhaps out of financial desperation or delusion. Few people really *expect* to win.

In 1992, however, an investment group in Melbourne, Australia, called the International Lotto Fund did expect to win a lottery with a multimillion-dollar prize. The group's organizers noticed that the Virginia state lottery had a game that involved picking six numbers from 1 to 44, and they observed that the chance of picking the correct six numbers was not bad as lotteries go. The total number of possible combinations of six numbers is given by 44!/(6!38!), the fraction part of the binomial formula. This number equals 7,059,052. Only a single combination would be the winning combination, so any particular combination of six picked at random had a 1/7,059,052 probability of winning.

The investment group set out to buy 7,059,052 lottery tickets in Virginia, one bearing each possible combination of six numbers. The purchase of a ticket with the randomly chosen winning combination would yield a prize of $27 million. The tickets cost $1.00 each. Thus, for an investment of $7,059,052 (not counting the administrative costs of purchasing 7,000,000 tickets—fairly substantial, I suppose), the return would be well over threefold, and the money would be secured very quickly and with little risk. The International Lotto Fund had the money to spend because it had the pooled resources of 2,500 small investors from Australia, New Zealand, the United States, and Europe, who each contributed an average of $3,000. Each $3,000 would yield $10,800.

The Virginia lottery winnings are not actually paid instantly, but as installments over 20 years. Still, the return would be $1.35 million per year over that period; an investor contributing $3,000 would collect $540 per year for 20 years. A *New York Times* article reporting the story on February 25 described the payoff as an investment from an accountant's perspective. Remember, no interest is paid during the 20-year payoff, and due to inflation each successive payment is worth a bit less than the one before; in addition, if you had the whole jackpot in hand at the time of winning, that capital could be used to earn income for the 20-year period. Nevertheless, the *Times* still considered it a fine investment, "equal to receiving a rate of return of about 16 percent on a 7 million dollar investment." Moreover, this was "a sure thing," whereas usually only the riskiest, most speculative investments pay rates of interest in the neighborhood of 16%.

The winnings would actually be slightly greater than that. In the Virginia lottery, there are also second, third, and fourth prizes for various other combinations, all of which would be present in the set of tickets held by the investors. But these are rather small additions to the jackpot. The second prize is $899 (there are hundreds of these, rather than a single such prize, since fewer numbers need be selected to win). The third prize is $51 (tens of thousands of combinations will do). The fourth prize is $1, just enough to allow you to buy your next chance of winning (hundreds of thousands of combinations qualify for this reimbursement). All told, having every possible ticket in hand would yield prizes totalling $27,918,561.

Buying every lottery combination and collecting the payoff looks like a sure thing, but there were three sources of risk that could threaten the profitability of the venture. One was not considered a realistic concern: lottery officials might decline to award the prize under the circumstances. The second potential problem caused greater worry because of its greater likelihood: others might pick the same winning numbers by chance, and the prize would be split among the holders of more than one ticket. At the time of the game in question, February 15, 1992, the lottery had been held 170 times. Most of the time, no one had won the jackpot; in fact, this had been the outcome 120 times. The jackpot had been awarded the remaining 50 times, but winnings had been shared 10 of those times. Sharing with one other winner and receiving an 8% return on investment would still be considered a tolerable rate of return compared to some other investments in 1992 (though not compared to the New York Stock Exchange at the time). A rate of return of roughly 5% would result if three tickets were sold with the winning numbers; this rate of return (or anything lower resulting from even wider sharing) is essentially a loss of money from the investors' perspective, because the $7 million could have been invested elsewhere at a higher rate of return.

The third possible danger to the profitability of the venture was that it might not be possible to complete the purchase of the tickets in a timely fashion. This would lead to the expenditure of a great deal of money for a large set of lottery tickets with insignificant prizes. The investment group therefore laid the groundwork carefully, making great efforts for the purchase to go smoothly. In advance of the purchases, they filled out 1.4 million slips by hand as required—each slip can be used to purchase five games (that is, sets of combinations). They had 72 hours to purchase tickets. Teams of people purchased tickets at 8 grocery store chains, for a total of 125 retail outlets. Grocery store employees involved in the purchase had to work in shifts to print the tickets at record speed. One grocery sold 75,000 tickets just in the 48 hours before the selection of the winning numbers. One chain of stores accommodated the purchase in such a way as to minimize the strain on its stores. The chain's corporate headquarters accepted bank checks for the sale of 2.4 million tickets, distributed the work of generating them among its stores, and had the tickets picked up by couriers.

As it turned out, the investment group's workers ran out of time. They were able to complete the purchase of only 5 million rather than 7 million tickets by the time of the drawing. But luck was with them: by the time ticket sales ended, they had a winning ticket and avoided splitting the prize, circumventing two major potential problems. It took them days to find the ticket, but they eventually presented it to lottery officials.

The officials were in a quandary, because it didn't seem fair to the public to try to win the game by buying all the combinations, rather than relying on chance. In addition, the public complained about being unable to purchase tickets because of the investment group's activities. The officials held hearings, at which one pizza deliveryman said, "No one wants to be in line behind anyone who's there for three or four days." There were reports of machines being labeled as "out of service" while they were being used to generate huge numbers of tickets for the bulk purchase, in the absence of the purchasers. Finally, it was officially declared that nothing was specifically wrong with purchasing millions of tickets to increase the chance of winning. However, the lottery commission promptly fashioned rules giving priority to those standing in line, over absentee buyers. Rules concerning maximum numbers of tickets to be sold at one outlet were also placed under consideration.

Lottery officials did pose a realistic threat to the investment, however. They balked at paying for another reason. They had found a single rule in place with the potential to invalidate the winning ticket. This was a state law requiring that the complete transaction take place on the premises where the machine prints the ticket. The intent of this rule had nothing to do with the sale of millions of tickets per se. Instead, it was meant to prevent middlemen from buying blocks of tickets and reselling them elsewhere. Some people might then buy tickets from the middlemen at a higher price than $1.00 in return for the convenience of not having to wait in line at the store. Such "scalping" of tickets is illegal because lottery tickets have a regulated price, and also because of the potential for counterfeiting. However, since some of the tickets were paid for at a company headquarters and printed at the company's grocery stores, state officials at first said they might refuse to pay. The grocery chain said it had never been made aware of the rule, and the attorney for the investors echoed this statement. In addition, the International Lotto Fund stated that it could not determine whether the winning ticket in its possession was one of those that had been paid for at headquarters and printed elsewhere. Finally, Virginia lottery officials announced on March 10 that the February 15 jackpot would be paid to the investment group, because of the threat of protracted legal wrangling and the uncertainty over where the winning ticket had actually been purchased.

Are there other ways to make certain or nearly certain that you can win the lottery? Studies have shown that despite people putting much credence in "runs" of numbers or tendencies for certain numbers to be selected in particular lotteries, these are never more than coincidences one would expect by chance alone. An article in the New York *Daily News* on November 8, 1996, purported to "reveal the secrets of Lotto." It was filled with statistics and not a word about the inherent fluctuations that could explain them. For example, it mentioned that the number 46 came up 22 times in the New York lottery, more than the next most common number (4, which came up 18 times). But only a year's worth of data were being considered, and no estimate was presented of the variability one would expect in proportions or frequencies observed in a series of this length. The article went on to say, "Some names have all the luck. Women named Mary, Maria, or similar variations have scored most frequently as winners. Among men, Josephs have hit the most jackpots." Is this luck? Aren't names like Mary and Maria among the most common names in New York, thus making their frequency among lottery winners simply an exact reflection of their proportion in the general population? Wouldn't "luck" imply that the probability of winning for a person named Mary is *greater* than we'd expect on the basis of proportions? Once again, the comparative method ought to be employed. Personally, I would think it more surprising, and more of an indication of special lucky attributes, if the most common first name for lottery winners were Esmeralda rather than Mary.

Miracles

One evening in 1950, a church choir in Beatrice, Nebraska, experienced both an annoying and a marvelous coincidence. The annoying coincidence was that everyone happened to be late for rehearsal that evening. This was pretty remarkable because the choir consisted of 15 people. It's not as if they were all delayed by the same snowstorm, for example. One delay involved an overly long nap. Another occurred because a car wouldn't start. Finishing up some geometry homework delayed someone else. The delays were not all perfectly independent, because some families contributed more than one choir member and delays were household specific. Still, there were a total of 10 households with delays that could clearly be considered independent.

What is the probability that 10 households will all experience delays and be late on a particular date? This is fairly ill defined, because we do not have good estimates of the chance of a single lateness. In a lottery, we can know the exact chance of a certain number or combination being selected. In many medical applications, we can rely on a wealth of published experience to provide estimates of the probability of responding to a particular therapy. Here, we can only guess, but let's say that the average choir member is late 10% of the time. Ten independent latenesses occur with the probability 0.1^{10}, which is equal to 1 in 10 billion. Even if lateness were more common, it would still be extremely rare for 10 sources of delay to occur in a single evening. If the chance of 1 delaying event in 1 household is 0.2, the probability of all 10 occurring is about 1 in 10 million.

The annoying coincidence was that everyone was late. However, it turned out to be a marvelous coincidence as well because the delays all occurred the very evening that a freak accident occurred. An explosion destroyed the church a few minutes after the time that choir practice was scheduled to start. In an article in the March 27, 1950, issue of *Life* magazine, choir members wondered aloud whether the peculiar series of delays might have been divine providence at work.

Very few churches end up destroyed, especially on a given evening when an extremely rare set of circumstances delays all choir members. What probability can we assign to the church's destruction? We need to multiply our probability of 10 delays, arbitrarily estimated as 1 in 10 million or 1 in 10 billion, by the probability of destruction in order to arrive at the joint probability that the delays would occur at that very point in time. Of course, the number we need is unobtainable (although I suppose insurance companies may have estimates), but it is bound to be a tiny probability. That said, it is noteworthy that Durham cathedral in England, which was built during the medieval period, was also destroyed in a freak storm. This occurred shortly after the Bishop of Durham had publicly questioned whether Mary was indeed a virgin. Are such events miraculous? Certainly the probabilities are vanishingly small.

Divine providence has often been invoked, even by statisticians, to explain exceedingly rare series of events. John Arbuthnot was a Scot and physician who found after medical school that he was more interested in statistics and mathematics than in medicine. In the late 1600s, he translated into English a book on probability written by the Dutch astronomer and physicist Christian Huygens. Arbuthnot added his own examples from games of chance. When this joint work was published, it was the first book in the field of probability to be printed in English. Thereafter, he published and taught widely in mathematics and became a fellow of the Royal Society in 1704. It was in the Royal Society's *Philosophical Transactions* that he published his "Argument for Divine Providence" in 1710.

Arbuthnot had noticed the small but consistent excess of male over female births mentioned in chapter 1. He looked at a series of data that consisted of 82 annual records of christenings from London and showed that the chance of having more males born than females each year for 82 years was minuscule. Moreover, he claimed, this had most likely happened for "Ages of Ages, and not only at London, but all over the world." Thus, he concluded, the chance of this phenomenon occurring by chance "will be near an infinitely small Quantity, at least less than any assignable Fraction. From whence it follows, that it is Art, not Chance, that governs."

Whose "Art"? And why govern the sex ratio at birth this way? Arbuthnot has this to say: "We must observe the external Accidents to which Males are subject (who must seek their Food with danger) do make a great havock of them, and that this loss exceeds far that of the other Sex, occasioned by Diseases incident to it, as Experience convinces us. To repair that Loss, provident Nature, by the disposal of its wise Creator, brings forth more Males than Females; and that in an almost constant proportion." It was in fact true, then as it is now, that males have higher mortality rates at every age than females do. This type of explanation of the sex ratio was also published 30 years later by the minister and demographer Johann Peter Sussmilch; his writing, based on statistical studies of Brandenberg and Prussia, was discussed in chapter 1.

These days, science tends to address immediate physical causes (the more immediate the better) rather than intermediate or ultimate causes. For instance, our explanation of the constancy of the sex ratio at birth focuses on how hormones are regulated and how they determine the sex of the child. Our control of medical and many other problems has advanced as a result of this kind of reductionist science, because the ability to alter immediate physical forces gives science great power when those alterations have observable, desirable effects. Consequently, the idea that something improbable might be proof of a distant providence is rather less widespread than formerly; the great rarity of a sequence of events does not, by itself, rule out mere coincidence or a physical cause. Moreover, it's hard to think of an event that would have *zero* probability by chance alone and would *have* to be divine in origin. On the other hand, perhaps there are occurrences so exceedingly rare that as a practical matter, we really should never have the opportunity to observe them by chance alone. I refer to observed events that might be, for example, as improbable as all the molecules of air moving to one side of the room. If we have only statistical reasoning to rely on, and no incontrovertible evidence as to underlying physical or divine causes, the source of the very rarest events must remain indeterminate.

5

The Waiting Game

Driving You Crazy

If you take the bus, you surely have had occasion to wonder why buses have to arrive in bunches after you've been impatiently waiting. Why *can't* they run at evenly spaced intervals? Intervals between trains also vary too much. And if you drive, you have similarly experienced (even in the absence of accidents) a sudden buildup of congestion that may dissipate as quickly as it started. Mathematics predicts this everyday experience. One branch, queuing theory, describes such phenomena as a group of commuters going to work and provides explanations of why they will almost never travel in a perfectly smooth and even flow.

Despite passengers' suspicions, the clustering of buses need not originate in the desire of the drivers to travel together in friendly packs. Suppose that the buses set out on their journeys at evenly spaced intervals, let's say 10 minutes apart. The trouble is, passengers do not begin waiting at evenly spaced intervals. There is random variability in how many passengers accumulate at the bus stop during those intervals. Even if they all intended to get to the bus stop at exactly the same time, one may oversleep, another get up unusually early, and so on. At some point, an interval will have a larger number of people than the average, and the bus that comes along at that time will have many people to pick up. Waiting for all those people to board and pay slows down the bus a bit. This delay narrows the interval between the crowded bus and the bus coming after it. Because he or she has been delayed, the driver of the crowded bus finds the crowding worsening progressively, with even more passengers waiting at the next stop than if the driver had arrived on time. The crowded bus gets slower and slower. Meanwhile, the bus behind the crowded bus becomes faster and closer, having fewer passengers to pick up than would typically accumulate during a full-length interval. As the journeys continue, the difference in ridership grows more pronounced, and the distance and time between the buses become shorter and shorter. Of course, the trend can be reversed. The bus traveling close behind, with fewer passengers, could suddenly have to pick up a very large number of people who started waiting in the very narrow interval after the crowded one passed, such as when lots of people leave a school or office at the same time. A book by Rob Eastaway and Jeremy Wyndham, *Why Do Buses Come in Threes?*, describes the above sequence of events and many other examples of queuing and clustering.

The flow of automobile traffic also "bunches up," whether city or highway driving is involved. On city streets, traffic lights let through a certain number of cars during the green interval. When traffic is moderate, the length of the green light is usually ample to allow the passage of the backlog of cars accumulated during the red light. Indeed, the timing of lights is usually set specifically to allow this to happen. However, since cars enter the system at random, on chance occasions there will be too many cars at the light to allow them all to pass in one green interval. Then, significant backups occur especially if the backlog reaches another light. Moreover, during rush hour such backups will be a typical occurrence, and even those lights that are adjusted to keep up with the flow of traffic may not help much. Another possible problem is that there may not be that much freedom to adjust the lights if a long green light holds back traffic flow at a major intersecting road.

It is perhaps more surprising that purely random variability in traffic density causes congestion on highways with no traffic lights. Yet, surely most drivers have had to slow down after traveling at full speed, on account of a slowly moving dense patch of traffic. Resignation sets in as you conclude that you are not going to make the trip as quickly as you had thought, but then the slowdown clears and the drivers in front of you return to the previous rate of speed. The even spacing of the cars is also restored. When you pass the spot where traffic had been clustered before, you see no evidence of any accident or other reason for the slowdown.

Here, too, traffic density is a culprit, for this phenomenon is not observed when traffic is light. When the roadway is saturated with cars, however, the tight spacing of the vehicles may be at the limit of drivers' toleration. Then, if a driver feels too close to the car in front, he taps the brake lightly to maintain a more comfortable distance. This may occur when new cars enter the highway up ahead, or because somebody slows down to avoid a bird, debris, or an irregularity in the road surface such as potholes. Once one driver touches the brake, the one behind him or her also feels too close and replicates the slowdown. The successive deceleration of a series of cars ends when it reaches a driver who, by chance, still has a satisfactory amount of "safety space" without slowing down, but in the meantime this deceleration has created a "bunching up" effect. Then a reversal occurs: the first car speeds up once it passes the initial cause of the slowdown (for example, an exit ramp), and the cars behind successively readjust their speed and spacing. To the more distant observer, it seems as if the effects of an accident have occurred, then vanished, and on passing the spot where the congestion had been present a few moments earlier, there is nothing to be seen that would provide a material explanation for the blocked traffic.

Some people hope to avoid queues by working at home and telecommuting. This reduces wear and tear on those individuals and vehicles that avoid travel, but you still can't escape queuing theory: its laws govern circuit requirements for things such as telephone calls and Internet access.

Historically, problems related to telephone circuits motivated the first mathematical treatment of queuing theory. In 1909, a Danish engineer by the name of Agner Krarup Erlang introduced the subject in the statistical literature. He was an employee of the Copenhagen telephone company and had to determine the optimal number of circuits (and operators to complete them, in those days) needed by the company in order to avoid severe congestion at peak call times. It became clear that there was a trade-off: circuits cost money, so the importance of making sure that calls could almost always be completed was balanced against a concern about building expensive excess capacity that would almost never be used. Consider the problem on a probabilistic basis. There's certainly never a *zero* probability of a given call volume at a given instant. But the most extreme call volumes are very unlikely (would everyone in the world really happen to call a particular neighborhood at the same moment?). It is not worth the effort and expense of building facilities to handle the most unlikely situations. Erlang realized that it was desirable to use a mathematical model to calculate a probability distribution of different levels of congestion given the nature and variability of phone call traffic. The typical waiting time to place a call and the typical size of the queue were also key system characteristics for which quantification was important, since they are measures of the adequacy of the system and are important components of customer satisfaction.

Analysis of Some Simple Queues

The equations that govern queuing phenomena have quite general applications. For example, in a nearby park there is a lovely lake and on fine days there are rowboats for rent. They can be rented at only one location, namely a shack in which the boats are stored. Customers enter, sign in, pay a deposit, and are given a life jacket and an explanation of safety rules. The boathouse worker then assists the customer in dragging the boat into the water. With no other customers in sight, it takes 5 minutes from the time a customer enters the boathouse to the time he or she is floating on the water. In the language of queuing theory, this 5-minute span is called the service time and is represented by s. The service *rate* is $1/s$, so in this example $s = 1/5 = 0.2$ customers served per minute, or 12 per hour.

The underlying service rate is assumed constant over time, although the sample of observations inevitably has some random variability associated with it. Some service times will be a bit longer or shorter than the next one. But, basically, they are samples governed by one constant underlying rate, and there is no sequential effect: all fluctuations are random, so what happened with a previous service time has no bearing on the next one. Therefore, we say that the service times may be assumed to be independent and identically distributed, in which the service rate $\mu = 12$ is a mean from an exponential distribution.

Of course, there's another part to figuring out the typical backup in the queue and the time to get through it: the arrival rate. This rate tells us how many new customers enter the queue per unit time. The Greek letter λ is often used to represent the arrival rate, and λ is often assumed to follow a Poisson distribution. Indeed, many real-world situations involve an arrival process for which the Poisson is an appropriate model: spans of time may often be divided realistically into intervals within which a single arrival would be usual, and additional arrivals possible but rare; very numerous arrivals would be exceedingly rare.

Suppose that we have a service time of 5 minutes, and an arrival on average every $7\frac{1}{2}$ minutes, that is, 8 arrivals per hour. We thus have $\mu = 12$ and $\lambda = 8$. How often is the boathouse worker busy? He or she can serve 12 persons per hour, but only 8 show up. Therefore, he or she is busy 8/12 or 2/3 or 67% of the time. This is the so-called utilization rate ρ. More generally, $\rho = \lambda/\mu$. The average overall waiting time in the system, including the service time, is $W = 1/(\mu - \lambda) = 1/(12-8)$; thus, the average waiting time is 0.25 hours, or 15 minutes. The average wait in the queue—pure waiting time, *exclusive* of service time—is given by $W_q = \rho W = \rho/(\mu - \lambda) = 0.67/(12 - 8) = 0.1675$ hour, or about 10 minutes. These figures seem counterintuitive: after all, the service time is shorter than the interval between arrivals, so why can't the boathouse worker finish with a customer before the next arrival? The reason is that there's no guarantee that the arrival will occur just when the boathouse worker has finished serving the previous customer and is ready to serve the next one. That's why the average wait must be longer than the average span of time between arrivals: arrivals have a 0.67 probability of arriving when the worker is helping someone else.

As these waiting times imply, there will be a backed-up queue even though the interarrival interval is longer than the service time. At a given moment, the average number of customers in the system is given by $L = \lambda/(\mu - \lambda) = 8/(12 - 8) = 2$. This includes the person whose service request is being worked on; the average length of the pure queue behind him or her in these circumstances is given by $L_q = \rho L = \rho\lambda/(\mu - \lambda) = 1.34$.

Suppose that it's an unusually beautiful spring day, and a weekend to boot. The service rate might remain unchanged, but the number arriving per hour is now 10 rather than 8. What do the queue and the waiting times look like now? μ remains equal to 12; with λ of 10, $\rho = 0.833$. At a given time, $L = 10/(12-10)$, so 5 people rather than 2 would be in the system. A high proportion of them would be in queue: $L_q = 0.833L$, or 4.165 on average (that is, 4 people would be waiting and the fifth would be in the course of being helped, with 16.5% of his or her service time remaining). The waiting time to get through the system is given by $W = 1/(\mu - \lambda) = 1/(12 - 10)$; thus, the average waiting time would be doubled, to half an hour. The average wait in the queue—pure waiting time, W_q—now would be 0.833(0.5 hour), which is 0.417 hour or 25 minutes, up from 10 minutes in the previous example.

What if 12 new customers arrived per hour? With both λ and μ equal to 12, their ratio ρ is equal to 1—the worker would be busy 100% of the time. The waiting times and queue lengths would all have denominators of 12 – 12, or 0; waiting times and queue lengths would be infinitely long, and if you joined the queue you would never get waited on. Sometimes this is a realistic model of a situation in which you actually find yourself! The solution here is to add additional servers. In fact, customer service–oriented businesses use models like this to examine in advance the conditions under which their systems would cease to function and to indicate the appropriate number of servers to avoid lengthy waiting lines or total system failure under varying assumed arrival rates and service times. "Servers" need not be limited to workers; the number of automated teller machines needed at various locations is determined in advance using the same logic, and numbers, as suggested by the equations. If ρ indicates that a server would be busy 100% of the time, then 2 would be required to keep the line moving; if ρ is 2 then the number of servers must be increased to 3, and so on.

Sometimes additional servers may be added until the average waiting time is quite reasonable, yet customers remain dissatisfied. This happens often in grocery stores. The lines at supermarkets are especially annoying because they usually are about the same length and you have to choose which cashier will check you out the fastest. I, for one, almost always pick the wrong one. No matter how it looks to me when I get in line, it is very rare that my line ends up finishing first. I see other people who started at comparable positions in their lines finishing before I do. Undoubtedly, you have had the same experience. Why does this happen?

Fancy queuing models are not needed to analyze this situation, just the simplest of probability reasoning. Suppose that there are 10 cashiers, equally proficient, and 10 lines of equal length from which to choose. The lines actually do tend to end up being of equal lengths—or, more accurately, of equal expected service times. That phrase reflects the fact that someone with a huge load of items tends to dissuade others from joining the line, until several customers with smaller loads accumulate elsewhere. When there are enough smaller loads to be processed by other cashiers' lines, people will again start to accumulate behind the one big customer. On an ongoing, steady-state basis, then, we can assume that the cashiers should take equally long to get to you.

Then why do we so often make the wrong decision about which line to pick? Think of it this way: even though there are 10 equal lines and 10 equal cashiers, only 1 can be the fastest. There is variability between the lines due to what are, at the outset, random unpredictable factors (such as a cashier needing more change, more bags, a price check, or a few moments to flirt with a customer). There are also random clusters of hard-to-scan items, and of fatigue and slowing down. Given these random factors, for our purposes it's as if 1 of the 10 lines is selected at random to be the winner; the probability must be 9/10 that it's *not* going to be the one you're in.

More Complicated Models

Of course, the models we've been discussing are simplified, and when additional factors are relevant they can be factored in. In the boathouse example, we have implicitly assumed that there is an infinite number of customers, an infinite area for them to wait around in, and an infinite number of available boats. Obviously, any real boathouse or lake would only be able to accommodate some finite number of boats, and, moreover, there would be an optimum number of boats. At a certain point, if the lake looks too crowded, renting a boat will start to look less desirable and the arrival rate will slacken. Queuing theory is often used to model waiting times in doctors' offices, too, and the same problem arises. The size of the waiting room is considerably less than infinite, and nonurgent patients may take one look at an overly crowded office and decide to go elsewhere (or be turned away by telephone when asking to be seen on a given day). Sophisticated queuing models often need to incorporate such feedback loops as these.

Also, the boathouse model relies on parameters reflecting the *average* values for distributions, but these result in a deceptive model if there is a lot of variability around the means. Suppose that service time is highly variable—perhaps "regular customers" at the boathouse actually require very little service time, while some newcomers have questions about acceptable means of payment, the prices for more than one person, or the fine print in a waiver of liability. There may also be nonrandom variation in service time, as the worker tires more at higher values of ρ (or toward the end of the day). Arrivals may be more clustered than has been assumed or may follow an entirely different, non-Poisson distribution.

Many very complicated queuing models are not mathematically tractable. Even when solutions to equations are not available, however, queuing models may yield results of great practical utility for business applications. Results are obtained by computer simulation, using the Monte Carlo method, named after the famous casino.

In Monte Carlo simulation, a computer generates random numbers that represent what happens to "make-believe" individuals in various situations. For example, a number might be generated at random from a Poisson distribution with a suitable mean, or from some other distribution that matches the observed data. This would provide the arrival time of a hypothetical person. A random selection from exponentially distributed numbers might then provide his or her service time. Data for the next person are generated using a new set of random numbers. This is done sequentially for a huge group of simulated individuals, numbering in the thousands or even millions. As each new person's data are generated, the computer enters the new arrival into the system, then has them "served," and calculates how many people would be in the queue up to that point. The computer keeps running totals as it adds successive simulated individuals and can handle millions of virtual arrivals and service times in a matter of seconds.

Although distributions' means are used to characterize arrival and service patterns, the simulated customers may bottleneck—or be few and far between for a while—as a result of patterns created by the randomness of the selection from the distribution. This is a better model of what actually occurs than can be provided by equations that do not take random variability into account. Also, simulation allows the calculation of the average backup as well as the distribution of the frequency of periods of unusual congestion. In addition, service time might be altered in the model to indicate changing conditions—it can be made longer after some specific number of customers has been served (to indicate a tired server) or shorter to model the speedier service resulting from additional servers. New arrivals may be stopped when the queue is longer than some number, in order to represent a limited waiting area; new customers can be added again once the number waiting diminishes. In long queues, the probability that a customer will abandon the queue in disgust may be specified and applied to those who are waiting. In short, the computer allows researchers to play with the so-called toy models they have created, to find realistic simulations and thereby practical solutions for real-world problems.

Queuing theory also helps companies make money. Using models provided by queuing theory, whether based on equations or simulation, businesses try to optimize the output of their customer service systems, in order to maximize profit and customer satisfaction. Several factors must be considered. For example, the server may be used at a utilization rate (ρ) between 0 and 1. Queue lengths and waits are low when ρ is closest to 0, because the probability is small that the server is busy when a customer arrives; conversely, as ρ gets close to 1, queue lengths and waits increase markedly. From the business's point of view, a ρ near 0 is not very desirable: customers will consider the service excellent (no waiting), but there is ordinarily a high cost for providing a service that would generate little revenue in return. A very high utilization rate may seem desirable because costly employees or ATMs don't sit around doing nothing, but customers will experience poor service and perhaps go to competitors; they may not join a long queue in any case and potential demand and hence revenue are lost.

The cost of service per customer varies inversely with the cost of a long queue. Combined costs are lowest where the two lines cross. The location of this optimum point is carefully modeled by large companies, which often have teams of people to study the distributions and parameters involved in their operations in order to simulate queues and estimate their costs. To this day, one of the primary uses of this type of analysis is in estimating the numbers of telephone circuits needed to serve a customer base, just as it was when Erlang initiated the mathematical analysis of queues.

Psychology in Queues

As my irritation in the grocery line implies, more than pure mathematics is necessary when using queuing theory. The psychology of the wait is an important part of determining acceptable queuing structures for businesses. Certain queues are rarely abandoned. For example, persons who have undressed, donned a gown, and are waiting to undergo a magnetic resonance imaging (MRI) scan rarely leave at that point. On the other hand, each additional minute waiting for the MRI is rated as much more unpleasant and unacceptable than a minute waiting for an ice cream cone or a turn at a rowboat.

Unexplained waits seem longer than waits for which an explanation has been offered, and waits whose expected duration is not specified are also perceived as longer and more irritating. In most urban commuter train systems, passengers consider recurrent minor unexplained delays a major irritant, even when the delays do not ultimately have much impact on time of arrival. Customer dissatisfaction over these objectively small nuisances can lead to loss of revenue, because other means of transportation are selected simply to avoid tension.

Business management is well aware that waits involving unoccupied time are more boring—and hence seem longer—than waits during which there is something to do. In New York at lunchtime, the waiting lines at banks are extremely long. To gain a competitive edge by improving the perception of its waiting times, the Manhattan Savings Bank once had a practice of providing live musical entertainment at lunchtime. This resulted in improved customer satisfaction. The crowded bank that I go to in New York has installed television screens showing the latest news, weather, and financial reports, so at least there's something interesting to look at while you wait; the waiting time is perceived as less of a waste because at least you've been catching up on the news. Similarly, many urban railway systems have added news screens on the platforms. This improves customer satisfaction without improving the actual waiting time (and serves as an additional source of revenue, as well).

One of the strangest examples of the psychology of queues is provided by the successful handling of customer complaints concerning long waits at the baggage carousel at Houston International Airport. The airport's management was constantly receiving a barrage of criticisms about the waits for baggage. These complaints pertained especially to flights arriving weekday mornings during the rush hour, laden with passengers on their way to business appointments (who were thus not inclined to tolerate poor service). The managers decided to pay a great deal of attention to the issue, because the criticism was so prevalent and so strident. They spent money increasing the number of baggage handlers and also sought out experts for advice on how to provide better service. After these measures, they wondered whether the typical customer experience was now acceptable. They had two forms of data to determine this. One was the average elapsed time from leaving the plane to having luggage in hand, which turned out to be 8 minutes, considered an acceptable time for luggage collection in the air transport industry. The other was the roster of customer complaints, which continued unabated.

The experts then took a new approach, reasoning that the actual waiting time involved was not the source of the problem but that some psychological aspect of the situation might explain it. When they considered the experience of the typical passenger with checked luggage more carefully, they noticed that the 8-minute time span was composed of 1 minute of walking from the plane to the carousel followed by 7 minutes of actually standing around next to it. Also, the carousel stood between the plane door and the taxi stand, enabling those waiting for checked luggage to see the others from their flight leaving immediately. A solution was implemented that did not change the total time from plane to taxi, but did change the psychology of the waiting time. Flights landing during the morning rush were brought in to gates at the opposite end of the airport from the baggage claim area. The walk from the plane to the carousel was now 6 minutes, while the time spent actually standing and waiting around for checked luggage—and envying those without it—was cut to 2 minutes. Customer dissatisfaction disappeared.

Sometimes, however, watching others while you are still waiting improves customer satisfaction rather than diminishing it. For example, amusement park designers make a point of having the lines for rides weave around the attractions, so that the view of those enjoying the ride provides an enticing advertisement of the fun that lies ahead. In this way, fewer people are induced to abandon the line out of impatience, and revenue is enhanced. In addition, Disneyworld has very long "crocodile lines," which weave in and out and keep people constantly moving. So, even though the waiting time is long, people are happier because they never have to stand still.

Amusement park lines usually don't go solely around the attraction; they also weave back and forth in an area so that you get a really strong impression of the number of people waiting. This, too, is often deliberate and plays on another aspect of customer psychology. Desire for a given ride is related in part to a customer's perception of how many other people want to go on it. This psychological effect is so strong that it persists even after the ride is over. Customers rate their enjoyment of a given ride lower on uncrowded days with no waiting, compared with the average enjoyment on days when there's a bit of a wait.

* * *

In all the models described thus far, a well-defined pathway has provided a template for calculation. Queuing theory allows estimates to be made that help us understand the effects of changing density on such phenomena as driving or customer service—but calculations must be done for a given queue, such as a certain bus route or service station. But what if the pathway is not specified in advance? How long will it take you to travel a given distance if the route you take is itself subject to random influences that determine not just its speed, but its very direction? A large body of mathematical theory has arisen to study such "random walks" and has applications in such diverse fields as economics and climatology. These random walks are the subject of the next chapter.

6 Stockbrokers and Climate Change

Paths Less Traveled

It's a whiteout! You left your campsite to hike and explore the canyon, and now you find yourself in a howling blizzard with no landmarks to guide you. The steel gray sky is mirrored in the snow underfoot, the color matching so perfectly that you can't see the horizon line. The view in every direction looks exactly the same. And it's becoming increasingly difficult even to look around, given the density of snow in front of your face. The fierce winds make it a struggle to stay upright, to keep walking—but you must stay upright, for to fall now is to be buried in the snow and die from the cold. So you keep walking, hoping for a chance encounter with the walls that enclose the canyon, because you know that there is a series of emergency shelters available at intervals along those canyon walls, and you will be able to reach safety. No canyon wall is more than 30 feet away, but traversing those 30 feet is not easy. You can't focus on a single direction because there is no way to know the right path to take, and anyway the interference of the wind makes you stumble along haphazardly: a step or two in one direction, half a step in another, then several steps in a row before reversing course again for a short distance. Your chance of taking a step in the right direction is just the same as taking a step backward away from it. You are tracing what scientists call a random walk.

A half-hour later, where will we find you? (In a shelter sipping tea, I hope.) We can't be sure exactly, but we *can* rate the probability of your reaching a given distance. For starters, we know that one direction is as good as another. So, if we mark the points that you've got an equal probability of reaching in half an hour, they will form a circle. If that circle shows distances having *large* probabilities, the radius is small; distances having *small* probabilities have larger radii. There is little likelihood of reaching distant zones.

It is easy to see that two extremes represent the least likely distances traveled. One is the worst case: no two steps will ever be lined up and you will stumble around and *only* retrace your steps over and over again. Thus, the distance traveled would be zero. The other extreme is that by random chance, every step will happen to be exactly in front of the other in an unbroken, completely straight line. Hence, the distance traveled would be maximized and equal to the number of steps taken times the average size of your steps. Just as in coin tossing, in which the first case corresponds to tossing heads and tails alternately for the entire half-hour, and the second to tossing only heads, these scenarios are extremely unlikely, but not impossible.

There is a likeliest distance, too, which also involves the numbers of steps and their average size. There is more to the calculation, because these are not all in the same direction. For estimation purposes, we can consider backtracking as creating a series of triangles crossing and recrossing previous pathways. Suppose that you manage 25 straight segments in the half-hour with an average length of 6 feet each. You don't travel $25 \times 6 = 150$ feet, but rather $\sqrt{25} \times 6$ feet (that is, $5 \times 6 = 30$ feet) because the straight-line distance is a giant hypotenuse cutting through all the overlapping triangles. Thus, the Pythagorean theorem can be used to determine how far you travel, and this provides the best estimate of the distance: 30 feet of unidirectional progress for 150 feet of walking, in this case.

Here's the mathematical rule: Imagine that you are about to take your ith step. Your displacement is x_i and you have an equal chance of taking a step to the left (distance $-d$) or right (distance d). On average, $\langle x_i \rangle = \frac{1}{2}(d) + \frac{1}{2}(-d) = 0$, so you get nowhere. But $\langle x_i^2 \rangle = \frac{1}{2}d^2 + \frac{1}{2}(-d)^2 = d^2$. Hence, after N steps you are on average a distance Nd^2 from where you began.

However, random walks can be used to analyze many more situations than walks in snowstorms. There are lots of phenomena in nature and in human behavior that follow random walks. One early application arose from the observations of a nineteenth-century Scottish clergyman and botanist, Robert Brown, who commented on the violent and seemingly random motions of spores that he had suspended in solutions in preparation for microscopic examination. They seemed to move as if kicked around, first this way then that, by some submicroscopic force. Similar motions were reported for microbes in water, whose paths had random components in addition to the direction seemingly intended by the germs. Such motions were also seen for particles of dust, smoke, or other very small materials suspended in air, and there, too, it seemed that there was an invisible force determining the paths of the particles.

There was a vigorous scientific debate at first about what Brown had reported and these other similar observations. Some people said that the motions of tiny particles were evidence of a life force in inanimate matter, which could at times give rise to living things through spontaneous generation. Many others said that the motions were simply the result of convection currents, and that the flow of heated liquid (or gas) was moving the tiny objects. Still others dismissed the convection arguments, countering with the fact that the movements of adjacent particles, whether spores, microbes, or dust, were uncorrelated. Thus, tiny currents were not the source of the motion, or there would have to have been at least several particles near each other moving the same way. Heat did have something to do with the motions, though, because cooling slowed these motions and made them less violent, numerous, and frequent, whereas heating had the opposite effect. Eventually it was realized that the motions were the results of collisions between molecules—always in random motion—and the tiny objects of study; it became clear that the motion was greater at higher temperatures because heat energy makes the molecules move around more often and with more energy. These movements became known as Brownian motion, and its physics and mathematics have been extensively studied. Although Albert Einstein is best known for his theory of relativity and won a Nobel Prize for work on the photoelectric effect, his 1905 article on Brownian motion is his most-cited scientific text. In fact, Brownian motion of suspended particles is a key piece of observational evidence in support of the existence of atoms.

Since this motion is completely random, the path of any one particle cannot be predicted, any more than the single path of one hiker in a blizzard can be predicted. What *can* be predicted from the mathematical nature of random walks is the *average* behavior of a system. We have seen an example in which random walk yields 30 feet of progress away from the center, but this "best estimate" is an average. This means that if you were so unfortunate as to have the "experiment" repeated many times, and you made an *X* on the ground at the end of each half-hour journey, the *X*s would form a ring around the monument with density greatest at 30 feet in diameter.

The larger the number of experiments or trials, the greater the certainty concerning the average outcome. Four or five wind-blown strolls may very well produce an average distance quite different from 30 feet. Four or five hundred would reduce the chance of deviation from 30, very markedly; four or five million, vastly more. So when the subject is molecules in a gas or liquid, rather than people, the average behavior of the system is very predictable indeed—after all, there are 6.02×10^{23} molecules in 1 liter of gas at standard atmospheric pressure and temperature—and is known as a set of Laws of Statistical Behavior. The attempt to recover the macroscopic laws of physics from the microscopic behavior of constituent atoms gave rise to the branch of physics known as statistical mechanics. Born in the nineteenth century, it remains vibrant today.

Laws of Statistical Behavior govern such phenomena as the mixing of solutions. If a strikingly colored solution is poured very gently into a large glass of plain water but left unstirred, random Brownian motion alone will cause diffusion of the color throughout the glass. The predicted ultimate outcome is a completely uniform color, and our experience is, of course, in accord with this statistical prediction. The mixing occurs because of the trillions of collisions occurring each second on the molecular level—so many that the chance of them being all in the same direction, and our expectation being thwarted, is negligible indeed. The "front" of the colored liquid moves along at a speed governed by the average distance traveled by each Brownian "kick" (a very tiny number), and the square root of the number of kicks (a very large number), in the same way that the hiker's walking distance is predicted. However, a very tiny number times a very large number yields a medium-sized number, so mixing of solutions as the result of Brownian motion alone is consequently a relatively slow process (which is why martinis have to be shaken or stirred but are never left to diffuse on their own).

Deviations from laws of statistical behavior are so improbable as to be of negligible concern in everyday life. You wouldn't expect to see millions of hikers walking, by chance stumbles, in precisely straight lines during snowstorms. Likewise, when a colorful liquid has thoroughly diffused through a clear one, it is considered to be at equilibrium. You would be shocked to see the mixing reversed on its own, so that the glass of liquid separates back into clear and colorful sections. The likelihood is minuscule that the average positions of the colored molecules will deviate from a uniform distribution throughout the water, because each downward thrust on a colored molecule is on average counterbalanced by an upward kick on a different molecule, close by. However, if all of the forces happened to be upward at a given moment, the color would move spontaneously from the bottom to the top of the glass, and the bottom would become clear.

To pick a potentially more alarming yet possible example, what if all the molecules of air in the room you are currently sitting in jiggled into a straight line and ended up on the other side of the room? To simplify thinking about this, imagine the room divided in half—the half you're in and the other half. We'll ignore distances and the time needed for molecules to travel away and just imagine the calculation this way: each molecule can be either on your side of the room or on the other. Let's say the joint probability of all *your* side's molecules jiggling by chance to the other half is 0.5^x, in which x is the number of molecules—many, many trillions. I don't know about your calculator, but on mine even $0.5^{1,000,000,000}$ rounds to 0, despite the use of scientific notation as a compact way of displaying numbers with 99 zeros after the decimal point. And $0.5^{1,000,000,000}$ is a much larger number than $0.5^{\text{many trillions}}$. Moreover, we haven't even allowed (more realistically) for the need to rule out molecules from the other half diffusing back to your half—ruling that out would make the probabilities even smaller.

The streets of New York City, dense with pedestrians at rush hour, provide a perfect example of the principle that collisions in large numbers can slow down travel time enormously, compared with the progress you can make at other times as a pedestrian without competition. Yet, my repeated experience of walking out from Grand Central Station in the course of many years' commuting pales in comparison with what a particle of light energy has to endure to exit the sun. The radius of the sun is approximately 420,000 miles, and at the speed of light a few seconds should suffice for a particle to traverse the sun and make an exit from it, no matter where in our star the particle was generated. However, so many collisions with atomic (and subatomic) particles occur that the average straight-line distance traveled in between collisions is roughly 1 centimeter at a time; then the particle is bounced off in some other direction. A random walk is followed, never an efficient means of transportation, and it averages hundreds of years (depending on starting location) before a light particle leaves the sun behind. It then takes a mere 8 minutes to reach Earth, on a straight-line path.

The Bucks Start Here

Let's come down to Earth for a moment and consider a practical problem that is of concern to most people: making money. You are not a light particle trying to exit the sun, but an individual trying to exit the workforce with an accumulation of enough money so that you have a comfortable retirement. Or perhaps you are even hard at work accumulating wealth so that you can leave early and rich (or have done so already). Many people pursue such goals by examining the stock market and attempting to invest in such a way that their investments outperform the market—that is, so that their stocks go up faster than the "basket" of stocks that gets averaged in such indices as the Dow Jones Industrial Average or the Financial Times stock exchange index. Every day in major newspapers and on the Internet, you can see graphs showing the recent behavior of stock market indicators. These vary from day to day, of course, but their general appearance is represented by the examples in figure 6.1. Much coverage in the media is devoted to the analysis of such numbers, including noting recent trends and correlations of those trends with particular recent news events or even just with recent changes in attitudes that are sometimes called "the psychology of the herd." The graphs are also examined for cyclical fluctuations, because business cycles might be an important determinant of future stock prices. In general, you would think that anything that helps you peg future stock prices correctly would help you make money: you would know when to buy something, when it will crest, and when to pull out.

If past history is any guide, the stock market really is a good investment. Stock market records go back more than a century, and if you pick any two dates 20 years apart you will find that a stock market investment (as measured by the Dow Jones Industrial Average, for example) outperformed "fixed" investments (by which I mean bonds, notes, and other financial instruments with a guaranteed but fixed interest rate). Stocks also outperform the price of gold and commodities.

Of course, there's no guarantee that past history is any guide. After all, the stock exchanges with long series of records are in countries that became the great industrial powers during that time span. Along with the unprecedented increase in expansion and prosperity came an unprecedented increase in the share values of companies. Nevertheless, amid all the bouncing around of prices within small spans of time, there is indeed a general upward trend in the value of securities traded in the market, and you might say that short-term fluctuations occur *around* an upward trend line, and that they are centered on that line. The trend seems likely to continue.

However, most people are interested in outperforming the market, in being *above* that line. Since an individual's financial interest lies

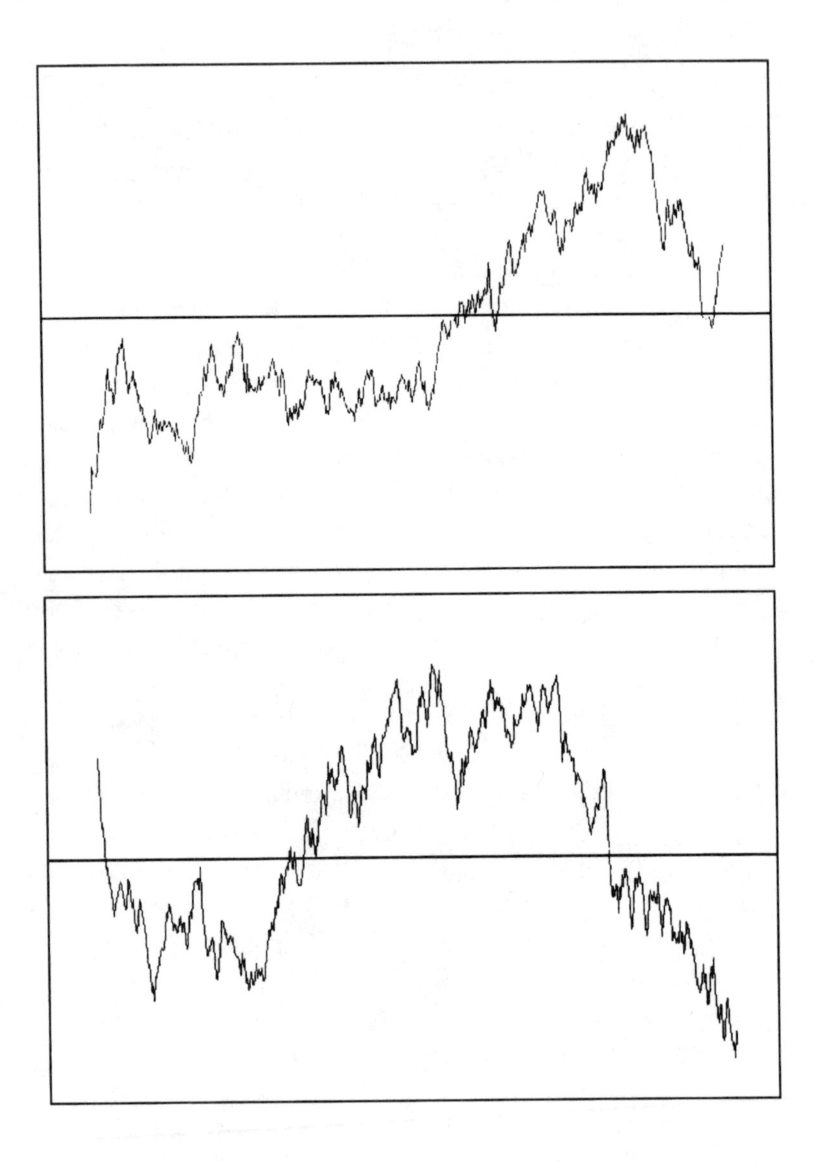

Figure 6.1. Two patterns of changes in a measurement over time: stock market price fluctuations or random walks?

in maximizing the rate of gain, few investors are content to buy the stocks in the Dow or the Financial Times "basket," go away for 35 years, and then retire with whatever they've got thanks to the general economic expansion of their era. Rather, people buy and sell constantly in order to gain and accumulate small advantages over time, and stockbrokers are there to advise on whether and when trades seem most advantageous. Even those whose stock market investments are limited to the pension plans provided by their employers are often given the opportunity to decide when to increase the stock market portion of their portfolio at the expense of fixed-income securities in order to take advantage of expected rises in the stocks. Some large institutions have the legal, fiduciary responsibility of investing the pension funds of thousands or even millions of people, in such a way as to maximize the return on investment; they, too, study trends and fluctuations in the market, because there is much competition to stand out as "better than average" at investing.

Suppose that the path of the Dow really is a random fluctuation about a general upward trend line. Even in the absence of any useful information from a stockbroker, an individual who invests in the Dow basket at a randomly selected date and sells the stocks at another randomly selected date has got a 50-50 chance of outperforming the market. Of course, there's also a 50-50 chance that the reverse is true. The thousands of such people who gain from their investments, whether they decided on them by using a Ouija board or a "tip" from a friend at work, would think that they had benefited from good advice. Those who took what amounted to random, informationless tips from similar sources and lost money would wonder what had gone wrong. And if stockbrokers gave advice (even random advice), half their customers would end up congratulating themselves on having retained such an insightful stockbroker, who invested their nest egg and made them so much better off.

Equity mutual funds are baskets of stocks picked by analysts and are supposed to offer an opportunity to outperform the market. At the end of a year, some of them do better than others, and those that do best get advertised widely, trumpeting their competitive advantage. The problem is that even among randomly chosen mutual funds, some would do better and some would do worse than average by chance alone. You could even have some fund doing better than most for a few years in a row and you couldn't be sure it wasn't a fluke, like five heads in a row when tossing a coin. However, about 20 years ago, 90% of equity mutual funds produced less profit than the stock market at large, as measured by the yardstick of the Standard and Poor's 500 stock index. Some of the failure to profit may have been due to service fees and charges, but some of it calls into question the fundamental utility of trying to predict the market. These days about 75% of mutual funds produce less profit than the S&P 500. The improvement may be real rather than an artifact or a fluctuation. While trading costs are generally lower now, there has been a proliferation of specialized mutual funds, and some focus on technical industries or the Internet. The existence of more numerous types of mutual funds, rather than a large number of homogeneous ones, makes it more likely that some—perhaps many—will tend to be focused on specific sectors of the market, such as technology or pharmaceutical stocks, that may be undergoing profitable expansion at the time; however, such trends are not necessarily sustained.

It's time to mention a feature of random walks that may seem surprising at first. If we look at a small local portion of a truly random walk, we will inevitably have a section that does not look random at all; in that section it may seem highly directional, or perhaps even cyclical. It would be as if we were to see a person making very good progress along a specific path. We may conclude, erroneously, that we are not observing a random walk, especially if that section is the only part of the data available for our inspection.

Suppose that you were told that figure 6.1 represents the prices of two particular stocks in recent months. At certain time points, would they have seemed good investments to you? Figure 6.1 actually illustrates random walk behavior seen in a coin toss series, not stock market prices. Every uptick in the trend line is a "heads" outcome, every downtick a "tails" outcome. The vertical axis indicates the running surplus of heads over tails. At the midway line halfway between the top and bottom parts of the figure, the two outcomes are equal; below that point, tails exceed heads while above it the reverse is true.

Remember, when the trend goes up past this line it does not indicate a huge number of heads "in a row"—just that the cumulative surplus is in that direction. Even with lots of intermixing of heads and tails, a surplus of one or the other can accumulate and become persistent. It happens that in these series there is some bouncing around, but a clear trend is evident: the proportion of heads is clearly going up in many places. Yet, figure 6.1 is actually an extract of small sections of a graph of many thousands of coin tosses. These are only local trends. The entire graph, which would run for miles, has almost exactly the same number of heads and tails, and the same number of upward trends as downward ones. Random walks do generate trends, and even cycles, but these can be misinterpreted. Especially in short runs of data, it simply may not even be possible to distinguish whether the data are simply from a random walk, or whether they are a graph of some directed activity or phenomenon that is taking place.

Burton Malkiel's best-selling book *A Random Walk Down Wall Street* examines the question: Does picking individual stocks by their past performance, or by analysis of their business potential, permit an investor to outperform a random selection of stocks? Malkiel points out that if thoughtful rather than random selection is to be useful, there must be a correlation between successive price changes for the same stock. Note that this successive correlation need not be perfect, but it must exceed what one would find by chance in random walks such as those depicted in figure 6.1. Careful selection of the right starting point in figure 6.1 can lead you to believe that the average percentage of heads in the coin tosses is increasing, so for a trend to be real it must be distinguishable from what would likely be seen on a random walk. Malkiel observes that successive price changes are not quite independent; they do tend to run in a given direction more than by chance, but this successive correlation is very weak. The conclusion for the investor is that the "buy and hold" strategy is best: let the general upward trend of the market increase the value of your holdings, and ignore particular trends in specific stocks (no matter what may lie behind the price changes), because it would be nearly impossible to distinguish the changes from random chance and, therefore, basically impossible to predict their future prices. The general up-trend is a much more reliable predictor of increasing value for your stocks.

Malkiel believes that "unexploited trading opportunities should not persist in any efficient market," or at least that the evidence is very weak for the existence of such opportunities. An efficient market, in this context, is one in which new information is reflected in prices as quickly as it becomes available. Prices move at random because they are affected most strongly by essentially unpredictable external news (such as wars, assassinations, political problems, and changes in fashions and recreation). Even events within a company have a heavy random component to them: a management approach that worked well in one period may be inappropriate in another, and managers may adjust or be replaced in time—or not. But what can the thoughtful approach to investing accomplish? Not much. As soon as relevant information becomes available, everyone's estimate of the impact of the news gets incorporated into the stock price through the impact of the information on supply and demand. As that estimate gets readjusted, the price comes to reflect the average assessment of the impact of the random news event. Therefore, the exploitation of information to get ahead of the average investor is no advantage, because everyone has the same information. Even the individual who wisely bucks the trend, dismisses a company's tale of woe, and ends up profiting from his or her optimism will have no better luck than random chance, next time. His or her decision to buck the trend was either haphazard (and not predictive of his or her next decision's success), or it was based on information (and his next decision will involve new information with new chances of random errors, uncertainties, and judgment error, the same as everyone else has to contend with). And there's no way to know whether the good decision was haphazard or the result of good judgment.

It seems that the way to get ahead is to have surefire insider information, but that is illegal because it is unfair and distorts the market; only if everyone has the same information will price changes reflect a stock's true value. Therefore, regulators want to see an efficient market with an honestly appropriate price for the stock. For example, in October 2000, new regulations began requiring American corporations to release any information with a potential bearing on their stock prices to all parties at the same time. Announcements of earnings and revenue projections and the granting of patents and the like cannot be presented to the cadre of Wall Street analysts before they are made to the general public. It used to be that you had to go to a stockbroker to get comprehensive information about a corporation, but that is a thing of the past. In the *New York Times*, October 20, 2000, U.S. Securities and Exchange Commission chairman Arthur Levitt expressed this view: "The kind of volatility, uncertainty, and even insider trading created by a system which depends on winks, nods, and whispers is far more dangerous to our markets than a system which respects the intelligence of investors who do not need intermediaries to interpret financial information for them."

At one time, it would have been difficult to require equal access to information. When Wellington defeated Napoleon at Waterloo, the news was sent by carrier pigeon to Nathan Rothschild, the wealthy investor in London who had arranged to receive it. Rothschild knew of the victory even before the British government did. It was obvious to him that British stock prices would increase wildly once the victory became common knowledge. Rothschild was a master investor and had a better strategy than to buy up stocks at once and profit from the coming run-up in value. He ran to the floor of the stock exchange and dumped everything he owned, putting all his stocks for sale at whatever price he could get. Other traders, seeing him react to a message in this fashion, assumed that Wellington had lost and dumped their stocks too. Prices plunged mightily, and only then did Rothschild buy up all the stocks he wanted, profiting doubly from both the sell-off he deliberately sparked and the inevitable giddiness in stock prices that he foresaw in the news of victory.

Even in the absence of market manipulation by a single individual, there are many examples of vast increases or decreases in wealth being created by "crowd psychology." After all, commodities such as ornamental diamonds or a particular artist's paintings command high prices only because people agree on (and convince others of) their value: the more people who create demand for an item, the more valuable it becomes. The introduction of tulips to the Netherlands from Turkey in the mid-1500s sparked the emergence of a speculative market for the flower bulbs. People began to bid on ever-rarer, ever-trendier varieties, of which only a small number of bulbs were available. In the early 1600s, single bulbs of rare, desirable types were so costly that they were exchanged for houses, used to purchase manufacturing plants such as a brewery, and offered as a bride's dowry. Often the bulbs did not even physically change hands; it was the title to a particular bulb, still in the ground, that was exchanged. The idea that prices could only rise became a self-fulfilling prophecy as it caused ever-greater demand. Returns on investments in tulips increased continuously and massively, cresting in the mid-1630s. The speculative bubble burst in 1637. A feeling that the run-up in prices would soon come to an end suddenly spread like a contagion and became a second self-fulfilling prophecy: the market almost overnight began to assess tulip bulbs as an investment of little value. Nothing objective had occurred, only the simple change in perception concerning the "value" of the tulip bulbs. Many families were bankrupted by the reversal, or lost homes that had been mortgaged for some flower bulbs. The value of tulips was so evanescent and so wholly dependent on a whim held in common that even today investment experts warn of "another tulip craze" when observing a speculative bubble expected to burst once the self-reinforcing desirability of an investment ceases.

In the short term, social forces such as belief in the continuing market value of tulips create nonrandom trends in prices: that is, for a time you can certainly invest more effectively by using your knowledge of the demand for tulips rather than by ignoring the information. In doing so, you will enjoy demonstrably better returns than you would if you invested randomly. The problem is that when viewed on a longer time scale, the investors' infatuation and subsequent disenchantment with such things as tulips is yet another random, unpredictable blip. You cannot predict what trendy new item investors will seek, nor when the craze will start or stop.

Self-perpetuating trends and beliefs may occasionally arise for seemingly the most analytical of reasons when a change in attitude concerns not a particular investment but beliefs about how a market can or should operate. For example, changes in regulations governing fair trading on stock markets are socially agreed upon and (in a sense) arbitrary rules; however, they may cause nonrandom trends, perhaps permanently altering markets. Changes in how markets operate may even come about simply because of the wide voluntary acceptance of novel economic concepts and theories. One example is the Black-Scholes options pricing theory. It started out as a mathematical model, which some thought accurately reflected the correct price to pay for stock options. Those who initially followed the theory made money from it, so others followed suit. Today the theory is so widely accepted that all decisions on the pricing of stock options are based on the Black-Scholes equation. It is considered correct, and, thus, it is a model that is "forced to be true" as a way of estimating the value of options: investors act as if it is in fact true and will only consider an option's price to be fair if it is the Black-Scholes price.

Let's be clear about what stock options are. A "call option" gives you the right to purchase a particular stock at a predetermined price. The option is a contractual right that you pay for in advance and that expires on a certain date. You do not have to exercise your right. However, suppose that stock A is selling at $200. You feel that stock A will rise and buy an option to purchase the stock at $220 two months hence. If the stock is in fact selling for less than $220 at that time, you have spent money on a worthless option. If the stock is selling at $250, the option "locking in" the price of $220 is worth $30 (regardless of what you paid for it). If the stock is worth $300 the option is worth $80, and so on.

Black-Scholes prices for call options are set by the following formula, devised by Myron Scholes, Robert Merton, and Fischer Black; the equation is considered so important that Scholes and Merton received the 1997 Nobel Prize in economics for it (by 1997, Black had died and Nobel Prizes are not awarded posthumously):

$$C = SN(d_1) - Le^{-rt}N(d_1 - \sigma\sqrt{T})$$

What is a fair price for an option to buy a stock? It is obvious that that price (C) has to depend on the initial price of the stock (S), the exercise price at which you can buy it later (L), and the time to expiration of the option (T). Two other factors come into play. One is the prevailing interest rate you could earn on a safe, fixed investment over the same duration; this needs to be included to allow for your potential earnings elsewhere, forfeited by the purchase of the option. Sometimes economists refer to this as "the time value of money." This time value is reflected by the compound interest term (in reverse): e^{-rt}. The final factor is the potential increase (or decrease) in the stock price, which is unknowable but allowed for by the inclusion of a measure of the stock price's volatility. The SD is used to measure this variability; σ represents specifically the SD of the continuously compounded rate of return on the stock, on an annualized basis.

Once these factors are all estimated, the previous equation is used to set the options price: it is a fraction of the stock's selling price, minus a fraction of the exercise price. The d_1 term is instrumental in setting this fraction. It involves previously estimated parameters and is given by

$$d_1 = \frac{\ln(S/E) + (r + \sigma^2/2)T}{\sigma\sqrt{T}}$$

Finally, $N(d_1)$ is included in the Black-Scholes equation because there is random variability in drawing a particular estimated d_1; d_1 is a sample observation from a normally distributed "universe" of possible d_1 observations. N indicates the probability of drawing it from a normal distribution (the normal distribution was discussed in chapter 2).

The equation for call options has become generally accepted; thus, there are massive nonrandom effects on markets as everyone sets the same price on options and they are not free to vary. The uniform behavior of the entire herd of investors to maximize profit in the same manner is clearly a deviation from pricing structures with random variability from investor to investor. (The same is true in principle for "put options," which give you the right to *sell* stocks at predetermined prices; corresponding equations exist for them.) Options prices are therefore now predictable and no one has an advantage in knowing them; they have been incorporated into the structure of how an investment market operates. No one gains from the situation by cornering an options market or by being able to invest with unique insight the way Rothschild could.

One night in January 1991, 176 years after Wellington's victory enriched Rothschild, the armed forces of the United States and her allies began to bomb Iraq. By the following morning, as Malkiel relates, there was a "clear indication that victory would be achieved quickly, [and] the Dow Jones industrial average opened 80 points over the previous day's close. The adjustment of market prices was immediate." Indeed, it seems to me that with modern electronic media and the advent of 24-hour on-line instant news and stock trading, stock prices are coming ever closer to being determined by the global, instantaneous inclusion of information freely available to all. This trend tends to equalize investors' judgments, limiting opportunities to exploit information and making it even more difficult to beat the average rates of return. There is a countervailing force, however. Only a few years ago, transaction costs were an important expense in large or frequent trades, but these costs have recently fallen. This leads to higher trading volume and perhaps a new source of volatility because there is little barrier to additional trades (and coupled with instantaneous news availability, more of a stimulus to make them). Anomalous chains of events could increase in number and create lots of new random winners and losers. Some people believe that an intelligent examination of instant news and stock prices will allow clever individual decision makers to beat the market thanks to their sustained good judgments. However, a representative sample of the outcomes of such individuals' behavior is not yet available to demonstrate definitively that a random walk can be avoided by investors in the long run.

Another theory also remains unproven in the context of stock prices. A body of mathematics called chaos theory has established that it is possible for a set of a few simple equations, involving a few factors as inputs, to generate output that is (to all appearances) completely random. The output of such chaotic processes satisfies all statistical tests for randomness, yet it is fully deterministic. Thus, when the equations are known, a given set of inputs corresponds to a particular set of outputs, making prediction possible. Some people hold that seemingly random fluctuations in stock prices actually follow specific equations and could therefore be predicted from current prices and other factors, but this is unproven. No one has specified the required equations and data and then successfully predicted the future course of stock prices.

A Heated Argument

The example of changing stock prices shows that it can be very difficult to distinguish between a random walk and a trend influenced by specific causes. Of course, there are many other examples of this problem including many with practical implications of various kinds. Concern that the Earth's climate is changing in ways that will prove troublesome for humanity has motivated studies of global warming. Thus, a great deal of what we might call "heated" argument centers on the issue of whether observed trends result from a random walk, or from some underlying process influenced by human activity, instead.

The Earth has recently gotten warmer. The trend is evident in data from the past 150 years and especially since 1900. Figure 6.2 shows single year changes and the same data displayed as 5-year moving averages. For a particular year, a 5-year moving average is that year's reading averaged together with the data from two adjacent years on either side. The use of moving averages is called a smoothing procedure. Since a given data point is used in several successive averages (along with other values), as the focus of the averaging moves along, the process dampens down isolated fluctuations and allows underlying trends to be displayed more clearly.

Surface temperatures are shown as differences in degrees centigrade from the average value in the center of the entire run of data. At first glance, these differences seem trifling, since they are completely contained within a range from roughly 0.3° below average to

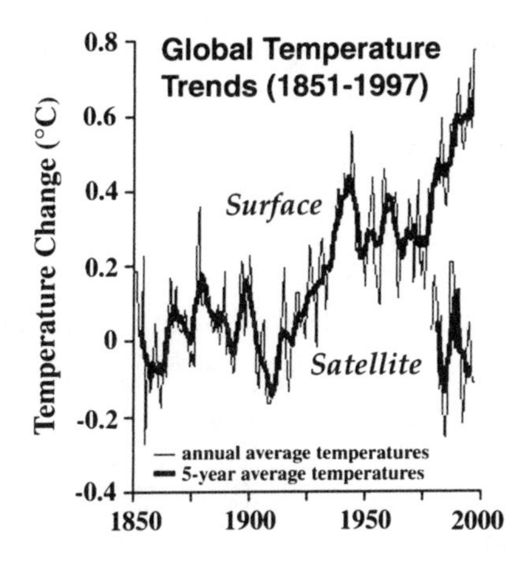

Figure 6.2. Global temperature trends. Measurements are relative to the average temperatures in 1851–70 (surface) and 1979–89 (satellite). *Sources:* Phil Jones, University of East Anglia (UK) and David Parker, United Kingdom Meteorological Office (surface data); John Christy, University of Alabama, and Roy Spencer, National Aeronautics and Space Administration (satellite data).

0.8° above average. However, vast amounts of energy are being retained by the Earth if the entire planet is being warmed by 0.5 or 1° within a century or so. The incorporation of such energy into the global climate system can unleash great storms and alter rain patterns, among other changes, especially if the warming is uneven and new patterns of convection currents arise in climatic systems. In addition, since figure 6.2 illustrates *average* changes, it obscures great inequalities in warming around the world. According to the U.S. Environmental Protection Agency's Web site, from 1951 to 1993 alone, average surface temperatures in cooler latitudes of North America have warmed up as much as 6°C while in other areas cooling of equal magnitude has been observed.

Incidentally, you can see that an immense amount of statistical data must be used by climatologists to arrive at these numbers. In thousands of locations around the world, weather station personnel collect data that (for consistency) are preferentially sought from stations that have been in operation for all or most of the 150-year span. For the surface *ocean* temperatures, millions of observations from ships during the relevant time span are used. The surface of the Earth is divided into a grid of 1° or less of latitude and longitude, and for a given year all the observations in a given box are averaged (whether land or sea temperatures). Then the boxes are all averaged to obtain an overall average for the Earth—after taking into account the fact that grid boxes may cover areas of different sizes.

The surface temperature data are thus combined, broad based, and reliable, and confidence in them is bolstered by consistency among adjacent weather stations and between land and sea measurements, and by the correspondence of local temperature data with local changes in glacier size, tree growth, and other phenomena.

The satellite data were compiled to study temperatures in the lower atmosphere. Although the trend in the data could be consistent with surface temperatures, so far it's a very short run of data with limited fluctuations and trend. Therefore, the data on the warming of the Earth's atmosphere are more controversial, whereas the data on global warming at the Earth's surface are rather clear-cut.

Let's accept that average global temperatures are rising. Why does global warming remain controversial? For one thing, there is an upward trend but it may be caused by nothing more than a random walk. After all, we saw graphs produced by coin tosses in figure 6.1, and they too had long stretches of time in which a persistent surplus of heads or tails arose from purely chance fluctuations, even though the underlying probability of 50-50 per toss remained unchanged, of course. There, too, our initial expectation might be that the time series of observations would show no trend and would fluctuate about the "zero line" because of the equal probability of heads and tails. Consequently, many scientists have made efforts to determine whether the global surface temperature series can be distinguished from a random walk.

In 1991, A. H. Gordon published a figure showing the sequences of cumulative changes in the direction of temperatures in the *Journal of Climate* (see figure 6.3). The top three graphs show the trends in the Earth's surface temperatures based on the yearly data, for the Earth as a whole and separately for two hemispheres. An upward movement for a particular year in any of these lines means that the temperature went up that year compared with the previous year in the series,

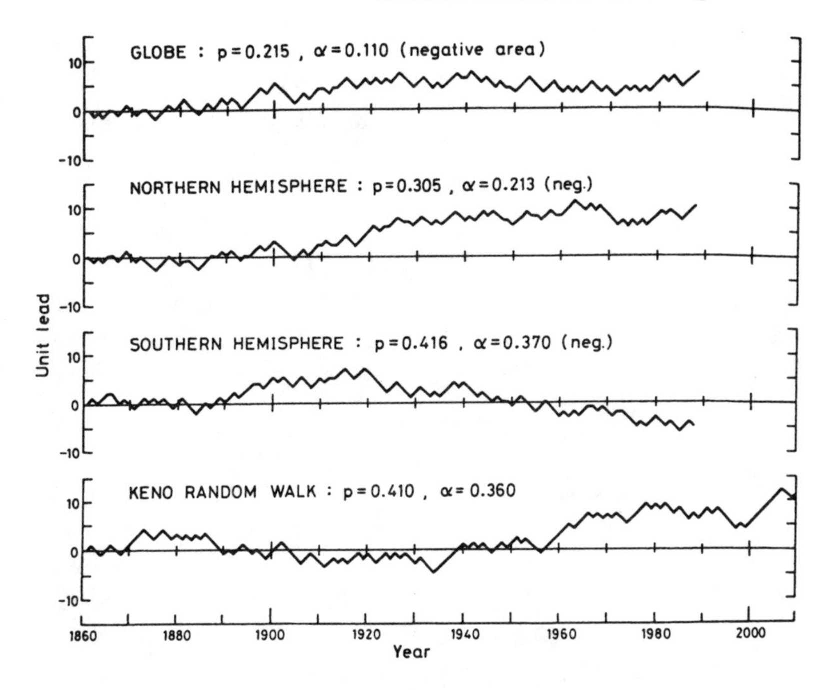

Figure 6.3. Sequences of cumulative changes in the direction of temperatures, globally and by hemisphere, with a random walk for comparison. Reproduced by permission of the American Meteorological Society.

and vice versa. The scale from −10 to 10 indicates the current surplus of "up years" or "down years" in the series. A mild (but by no means uniform) tendency for temperatures to be higher in successive years results in a rather persistent stay in the zone above the zero line in the overall (top) graph, and for the Northern Hemisphere as well. In the Southern Hemisphere, the line remains in positive territory for a while, then turns consistently negative. The bottom graph shows the results of a random casino game called Keno, as a series of outcomes "measured" simply as odd or even numbers. As with temperature change, the scale indicates the current surplus of odds or evens.

We might expect that the proportion of time spent above or below the "zero axis" in graphs such as these would be one-half. None of the four series spend their length divided so evenly. In fact, random walks tend to drift away from the line of even distribution, and the random walk of the Keno game here is no exception; the α value is 0.36, which indicates that 36% of the Keno walk is spent on one side of the line (that is, below it). The probability (p) of such an imbalance is fairly large, 0.41, so this is not a surprising finding. Southern Hemisphere temperature data have a similar α (0.37) and p (0.416) and are indistinguishable from the results of a game of chance. The sequence shown for the Northern Hemisphere is not much better, with $\alpha = 0.213$. The p value shows that between 30 and 31% (nearly one third) of random walk sequences of this length will spend 21.3%, or more than one fifth of their time, in the negative region.

The biggest difference from the Keno random walk is seen in the combined global data, in which 11% of observation points are in the negative zone. This means that in 89% of yearly observations, the temperature remained above average (regardless of whether it went up or continued back down in the direction of that average). Yet, the probability of getting the 11/89 split in a series this long is not *too* far-fetched: it's 0.215, so more than one-fifth of random walks this length would have this type of persistent upward trend.

This may seem surprising, but we are looking at the series of random independent tosses in a special way: we are not just counting the heads and tails or ups and downs, but whether the cumulative sequence of heads or tails at a given point has got heads or tails in the lead. If all the heads come first, then all the tails, you would have 50% of each as expected—in the form of a long run-up followed by a long run-down. Of course, that's an extreme example, but once a rare, very positive stretch occurs, it is likely to be a long time until an equally rare correspondingly negative stretch occurs. Thus, surpluses (once they have started) tend to persist in random walks.

A Short, Frank Comment about Wiener Processes

You may have noticed that temperature fluctuation is continuous, while Gordon simplified matters by reducing it to a discrete step that is either upward or downward. The mathematics is thus made similar to binomial sequences, providing an interpretation that is suitable to our purposes: either there is a tendency upward or there is not, irrespective of the magnitude. This analysis properly involves a random walk, because a random walk has steps. There is an analogous mathematical formulation called Wiener processes, comparable in concept to random walks but used for examining continuous fluctuations. They are named after Norbert Wiener, an American mathematician who studied processes such as Brownian motion that are dependent on random events. He was very active in World War II, applying his talents to optimization of antiaircraft gunnery placement, to filtering noise from radar signals, and to the devising of coding machines. Sometimes Wiener processes rather than random walks are used to model global warming, but this is relatively rare; in the case of climate change, the annual step size is tiny, and the first question of interest is in fact a simplified one: Is the Earth warming or not?

Dead Seas and Tropical Flora

In any case, Gordon concluded that the global temperature series could not be distinguished from a random walk. Four years later, Wayne A. Woodward and H. L. Gray published in the *Journal of Climate* an important addition to the examination of these data. They examined several possible models for the global temperature series. One was a linear trend through the points

$$\text{temperature at time } t = \alpha + \beta t + z_t$$

in which α and β are constants. α is the line's intercept, and β is the slope showing how steeply the trend is occurring per unit of time. The term z_t allows the addition or subtraction of a random tendency around the line. Another model was the same with the addition of a quadratic (squared) term, to allow the underlying line to be curved rather than straight:

$$\text{temperature at time } t = \alpha + \beta_1 t + \beta_2 t^2 + z_t$$

Finally, they looked at the results of an autoregressive model—one in which no specific preset formula for a line is used, neither straight nor curved. Instead, a moving average is used. Each successive data point is predicted from the moving average so far, plus a little bit of a trend. The trend is randomly chosen but is more likely to be upward if the last change was upward, and more likely to be downward if the reverse is the case (as if we had weighted rather than fair coins to toss). The global temperature time series predicted by this third model provided the best match to the actual values.

The implication is that the global temperature series has random trends rather than being the result of an inexorable fixed process that could be well described by a deterministic equation. The temperature series is more characteristic of data with randomly started yet persistent trends. They bounce around much more than would occur in a predetermined change (plus or minus some random noise) each year; yet, the bouncing's direction is significantly and highly dependent on the immediately previous trend. Thus, Woodward and Gray comment that the current observed trend may abate in the future, adding that "the results shown here also suggest that if only 30 or 40 more years of reliable temperature data (or its equivalent) were available, then a more definitive conclusion could be made concerning whether the trend should be forecast to continue." Those who believe that "greenhouse gases" generated by human activity are trapping heat, confident that current trends will continue, feel that it will soon be amply demonstrated that global warming is being observed.

This will be important to follow up, for much is unknown about global warming. Should we really consider the change depicted in figure 6.3 as a striking change? In the Cretaceous period, the Earth was so warm that what we term tropical flora flourished everywhere. Ice ages come and go on Earth, for reasons not yet fully understood. Among the factors involved in such massive shifts are continental drift, prevalence of volcanic or other dusts, and prevalence and reflectivity of various types of clouds and ice sheets. There are also changes in the shape of the Earth's orbit and in the tilt of the Earth that have an impact on climate change. For example, periodic changes in the elliptical shape of the Earth's orbit, and in the tilt of the Earth, influence the solar energy received. In fact, the Mediterranean was a desert between

5 and 6 million years ago, as can be confirmed by the types of fossils and the minerals formed then (gypsum and halite crystals) and still present under the sea floor. The flow from the Atlantic through the Strait of Gibraltar was impeded by geological processes, and the evaporation rate exceeded the replenishment of water from precipitation and rivers. One reason for the high evaporation rate was the tilt angle of the Earth's orbit around the sun at that time, which produced unusually high levels of heat energy. A 1999 article in *Nature* by Wout Krijgsman notes that the "drier Mediterranean climate" was the reason that "for a half million years, scattered bodies of water more saline than the Dead Sea retreated across the isolated, salt-encrusted Mediterranean basin."

So is the trend seen in the past century and a half "man-made"—is it more than what we would see because of natural fluctuation? Perhaps we had best compare this recent experience with the immediately prior centuries, in order to tease out the effects of human activity against a backdrop of relative stability. Data for this period cannot be obtained from weather stations, of course, but can be inferred from the types of plants evidenced by fossil pollen, and from the speed of growth manifest in tree rings. There are many other sources of information as well. Deep-sea sediments contain bits of seashells, composed principally of calcium carbonate ($CaCO_3$). The oxygen in this compound may be of several isotopes. Seawater is subject to a temperature-dependent evaporation process, and heavier isotopes of oxygen are more common in seashells when it is cold. The heavier form of the seawater is evaporated less readily. The temperature of seawater in the past can therefore be determined from the ratio of the isotopes in seashells at various depths in cores brought up from the bottom of the ocean. A similar isotopic fractionation of oxygen from bubbles in cores of ice found in glaciers at various depths permits an assessment of historic and prehistoric air temperatures. The increasing presence of methane, carbon dioxide, and other gases associated with human activity—sometimes called "greenhouse gases"—is also documented in the sequence of bubbles obtained from glacial cores.

All these sources of data provide convergent evidence, strengthening the conclusion that during the last thousand years in the Northern Hemisphere the twentieth century was by far the warmest. Indeed, in a thousand years the warmest four were 1990, 1995, 1997, and 1998. Yet, some people note that there was a "little ice age" in the Middle Ages and believe that the turnaround is merely part of the usual cycle of recovery. The speed and ultimate magnitude of the upward trend will soon permit a more definitive answer to whether global warming is the result of human pollution of the Earth with greenhouse gases or simply part of natural cycles that would occur in any event.

Additional controversy surrounds the question of the *effects* of global warming if the trend should continue. Will glaciers in Greenland, Antarctica, and elsewhere melt, causing the sea level to rise? Will new convection patterns in the oceans change sea life and affect food sources, while new patterns in the atmosphere change rainfall and also affect the agricultural productivity of various regions for better or worse? The Earth's climate system is vastly complex; it is uncertain what parameters would be needed to make predictions, and even more uncertain what the actual values of all the many parameters are. There are undoubtedly feedback loops that may minimize or accentuate trends. The problems in learning how the Earth's climate works, what is influencing it, and where it is headed are reminiscent of the problems encountered in making predictions about the course of epidemics that were discussed in chapter 1.

Perfect Models

At one time there was a faith that equations, provided with accurately estimated parameters, could be used to predict all physical systems. That was during the Enlightenment, when educated people in Western society were agog with the discovery that many natural phenomena can indeed be predicted by mathematical functions. The late 1600s and early 1700s saw triumphs of understanding and mathematical prediction such as the publication of Newton's *Principia Mathematica*, the elucidation of relationships such as the inverse-square law of planetary force, and the prediction of planetary orbits and the trajectories of comets. The extraordinary success of this style of thinking went on for a century and more, leading the French mathematician

Pierre-Simon Laplace to make this observation in the early 1800s: "An intelligence that, at a given instant, could comprehend all the forces by which nature is animated and the respective situation of the beings that make it up, if moreover it were vast enough to submit these data to analysis, would encompass in the same formula the movements of the greatest bodies of the universe and those of the lightest atoms. For such an intelligence nothing would be uncertain, and the future, like the past, would be open to its eyes."

This is a statement of faith that equations govern all physical events, and it is also an interesting speculation that you could use equations to predict the course of all future events from the past, and project them backward as well to know how current events had originated. However, note that such omniscience is implicitly forever remote from our experience: you would need every relevant equation and perfect knowledge of every necessary parameter to make predictions. Current thinking provides a realistic counterweight to Laplace's flight of fancy. You could never know all you would need to estimate *perfectly* the behavior of a physical system like the Earth's climate. For example, to obtain all the necessary data about temperature that perfect prediction requires, we would have to cover the Earth's surface and atmosphere entirely with thermometers. That, of course, would be impossible and would also change the Earth's reflectivity and hence its climate. Thus, there is no possibility of perfect knowledge of the Earth's temperature in any of the years in figure 6.3. We introduce statistical uncertainty to even the most perfect model. And that uncertainty exists for each of the many parameters that one would need to model climate. Moreover, how would we submit such vast amounts of information to analysis? Some people have estimated that a computer big enough to model the Earth's climate with extreme accuracy would have to be so vast that all the material in the universe would be needed to fashion it. After all, it currently takes about 24 hours of computer time for the most accurate predictions of 36 hours of weather—and they are hardly infallible predictions. While the need for a universe-sized computer may be an overstatement (particularly if quantum computing ever becomes practical), the point is that Laplace's dream seems to be impossible in principle.

Sir Robert May, F.R.S., is an outstanding scientist, a professor at Oxford, and sometime chief scientific advisor to the prime minister of Great Britain. Originally trained in Sydney as a physicist, he now specializes in the application of mathematical models in ecology and epidemiology. Yet, May (a former mentor of mine, whose very name indicates uncertainty) has observed that a statistical model must be a "caricature of reality": it is a recognizable representation of what really happens, but it is simplified and restricted to a few key (if distorted) features. Nevertheless, models are an important aid to human judgment, and our future depends in part on sensible application of statistical models to everything from drug development to climatology. Knowledge of the role that probability and chance play in life are important elements of deciding on courses of action. If we add to this mix the application of wisdom and judgment, then we perhaps will have a better future—or at least make it more probable.

Index